CHEMISTRY RESEARCH AND AP

INFORMATION ORIGINS OF THE CHEMICAL BOND

Chemistry Research and Applications Series

Applied Electrochemistry
Vijay G. Singh (Editor)
2009. ISBN: 978-1-60876-208-8

Handbook on Mass Spectrometry: Instrumentation, Data and Analysis, and Applications
J. K. Lang (Editor)
2009. ISBN: 978-1-60741-580-0

Handbook of Inorganic Chemistry Research
Desiree A. Morrison (Editor)
2010. ISBN: 978-1-61668-010-7

Handbook of Inorganic Chemistry Research
Desiree A. Morrison (Editor)
2010. ISBN: 978-1-61668-712-0 (E-Book)

Solid State Electrochemistry
Thomas G. Willard (Editor)
2010. ISBN: 978-1-60876-429-7

Mathematical Chemistry
W. I. Hong (Editor)
2010. ISBN: 978-1-60876-894-3

Mathematical Chemistry
W. I. Hong (Editor)
2010. ISBN: 978-1-61668-440-2 (E-Book)

Physical Organic Chemistry: New Developments
Karl T. Burley (Editor)
2010. ISBN: 978-1-61668-435-8

Physical Organic Chemistry: New Developments
Karl T. Burley (Editor)
2010. ISBN: 978-1-61668-469-3 (E- Book)

Chemical Sensors: Properties, Performance and Applications
Ronald V. Harrison (Editor)
2010. ISBN: 978-1-60741-897-9

Macrocyclic Chemistry: New Research Developments
Dániel W. Fitzpatrick and Henry J. Ulrich (Editors)
2010. ISBN: 978-1-60876-896-7

Electrolysis: Theory, Types and Applications
Shing Kuai and Ji Meng (Editors)
2010. ISBN: 978-1-60876-619-2

Energetic Materials: Chemistry, Hazards and Environmental Aspects
Jake R. Howell and Timothy E. Fletcher (Editors)
2010. ISBN: 978-1-60876-267-5

Chemical Crystallography
Bryan L. Connelly (Editor)
2010. ISBN: 978-1-60876-281-1

Chemical Crystallography
Bryan L. Connelly (Editor)
2010. ISBN: 978-1-61668-513-3 (E-Book)

Heterocyclic Compounds: Synthesis, Properties and Applications
Kristian Nylund and Peder Johansson (Editors)
2010. ISBN: 978-1-60876-368-9

Influence of the Solvents on Some Radical Reactions
Gennady E. Zaikov, Roman G. Makitra, Galina G. Midyana and Lyubov.I Bazylyak
2010. ISBN: 978-1-60876-635-2

Dzhemilev Reaction in Organic and Organometallic Synthesis
Vladimir A.D'yakonov
2010. ISBN: 978-1-60876-683-3

Analytical Chemistry of Cadmium: Sample Pre-Treatment and Determination Methods
Antonio Moreda-Piñeiro and Jorge Moreda-Piñeiro
2010. ISBN: 978-1-60876-808-0

Binary Aqueous and CO2 Containing Mixtures and the Krichevskii Parameter
Aziz I. Abdulagatov, Ilmutdin M. Abdulagatov and Gennadii V. Stepanov
2010. ISBN: 978-1-60876-990-2

Advances in Adsorption Technology
Bidyut Baran Saha and Kim Choon Ng (Editors)
2010. ISBN: 978-1-60876-833-2

Electrochemical Oxidation and Corrosion of Metals
Elena P. Grishina and Andrew V. Noskov
2010. ISBN: 978-1-61668-329-0

Electrochemical Oxidation and Corrosion of Metals
Elena P. Grishina and Andrew V. Noskov
2010. ISBN: 978-1-61668-329-0 (E-Book)

Modification and Preparation of Membrane in Supercritical Carbon Dioxide
Guang-Ming Qiu, Rui Tian, Yang Qiu and You-Yi Xu
2010. ISBN: 978-1-60876-905-6

Thermostable Polycyanurates: Synthesis, Modification, Structure and Properties
Alexander Fainleib (Editor)
2010. ISBN: 978-1-60876-907-0

Combustion Synthesis of Advanced Materials
B. B. Khina
2010. ISBN: 978-1-60876-977-3

Structure and Properties of Particulate-Filled Polymer Composites: The Fractal Analysis
G. V. Kozlov, Y. G. Yanovskii and G. E. Zaikov
2010. ISBN: 978-1-60876-999-5

Information Origins of the Chemical Bond
Roman F. Nalewajski
2010. ISBN: 978-1-61668-305-4

Cyclic β-Keto Esters: Synthesis and Reactions
M.A. Metwally and E. G. Sadek
2010. ISBN: 978-1-61668-282-8

Wet Electrochemical Detection of Organic Impurities
F. Manea, C. Radovan, S. Picken and J. Schoonman
2010. ISBN: 978-1-61668-661-1

Wet Electrochemical Detection of Organic Impurities
F. Manea, C. Radovan, S. Picken and J. Schoonman
2010. ISBN: 978-1-61668-491-4 (E-Book)

Quantum Frontiers of Atoms and Molecules
Mihai V. Putz (Editor)
2010.ISBN: 978-1-61668-158-6

Molecular Symmetry and Fuzzy Symmetry
Xuezhuang Zhao
2010. ISBN: 978-1-61668-528-7

Molecular Symmetry and Fuzzy Symmetry
Xuezhuang Zhao
2010. ISBN: 978-1-61668-375-7 (E-Book)

Molybdenum and Tungsten Cofactor Model Chemistry
Carola Schulzke and Prinson P. Samuel
2010. ISBN: 978-1-61668-750-2

Molybdenum and Tungsten Cofactor Model Chemistry
Carola Schulzke and Prinson P. Samuel
2010. ISBN: 978-1-61668-828-8 (E-Book)

Rock Chemistry
Basilio Macias and Fidel Guajardo (Editors)
2010. ISBN: 978-1-60876-563-8

Boron Hydrides, High Potential Hydrogen Storage Materials
Umit B. Demirci and Philippe Miele (Editors)
2010. ISBN: 978-1-61668--362-7

Tetraazacyclotetradecane Species as Models of the Polyazacrown Macrocycles
Ryszard B. Nazarski
2010. ISBN: 978-1-61668-487-7

Chemical Reactions in Gas, Liquid and Solid Phases: Synthesis, Properties and Application
G.E. Zaikov and R.M. Kozlowski
2010. ISBN: 978-1-61668-671-0

CHEMISTRY RESEARCH AND APPLICATIONS SERIES

INFORMATION ORIGINS OF THE CHEMICAL BOND

ROMAN F. NALEWAJSKI

DEPARTMENT OF THEORETICAL CHEMISTRY
JAGIELLONIAN UNIVERSITY
CRACOW, POLAND

Nova Science Publishers, Inc.
New York

For permission to use material from this book please contact us:
Telephone 631-231-7269; Fax 631-231-8175
Web Site: http://www.novapublishers.com

LIBRARY OF CONGRESS CATALOGING-IN-PUBLICATION DATA

Nalewajski, R. F. (Roman F.)
Information origins of the chemical bond / Roman F. Nalewajski.
p. cm.
Includes bibliographical references and index.
ISBN 978-1-61668-305-4 (softcover)
1. Chemical bonds. I. Title.
QD461.N269 2009
541'.224--dc22

2009052735

Published by Nova Science Publishers, Inc. ✦ New York

CONTENTS

ACRONYMS

AIM	Atoms-in-Molecules
AO	Atomic Orbital(s)
a.u.	Atomic units
BEF	Binary Entropy Function
BO	Born-Oppenheimer (adiabatic), approximation
CBO	Charge–and–Bond-Order (density), matrix
CG	Contra-Gradience, criterion
CI	Configuration Interaction, theory
CS	Charge Sensitivities
CT	Charge Transfer, between AIM
CTCB	Communication Theory of the Chemical Bond
DA	Donor-Acceptor, complex
DFT	Density Functional Theory
ELF	Electron Localization Function
EPI	Extreme Physical Information, principle
F	Fisher, information
HF	Hartree-Fock, theory
HK	Hohenberg-Kohn, DFT
HSAB	Hard and Soft Acids and Bases, principle
IT	Information Theory
KL	Kullback-Leibler, entropy-deficiency
KS	Kohn-Sham, method
LDA	Local Density Approximation, of KS theory
ME	Maximum Entropy, principle
MED	Minimum Entropy Deficiency, rule
MO	Molecular Orbital(s), theory

OAO	Orthogonalized AO
OCT	Orbital Communication Theory
P	Polarization, of AIM
PES	Potential Energy Surface
RHF	(Spin) Restricted HF, theory
S	Shannon, entropy (information)
SBC	Symmetric Binary Channel
SCF LCAO MO	Self-Consistent-Field–Linear-Combinations-of-Atomic-Orbitals–Molecular-Orbital, theory

PREFACE

The *Information Theory* (IT) [1-6] is one of the youngest branches of the applied probability theory, in which the probability ideas have been introduced into the field of communication, control, and data processing. It has originated from the needs of practice, to create a theoretical model for a transmission of information, and evolved into an important chapter of the general theory of probability. Its foundations have been laid in 1920s by Sir R. A. Fisher [1], in his classical measurement theory, and in 1940s by C. E. Shannon [3], in his mathematical theory of communication.

An understanding of the distribution of information in molecules and its displacements accompanying chemical reactions, which involve the bond-forming and/or bond-breaking processes, provides an alternative perspective on molecular electronic structure. An insight into the entropic origins of chemical bonds and their coupling in chemical phenomena is central to many branches of chemistry [7]. The electronic quantum-mechanical state of a molecule is determined by the system wave function, the (complex) amplitude of the particle probability distribution which carries the information. It is thus intriguing to explore the information content of the electronic probability distribution in a molecule, and to extract from it the pattern of chemical bonds, trends in chemical reactivity, and other molecular descriptors, e.g., bond multiplicities ("orders") and their covalent and ionic components.

In particular, it is of great interest to examine the identity of bonded atoms in molecules, the exchanges of information (communications) between them, as well as the information representation of subtle electron redistributions in chemical processes, e.g., those accompanying formation or breaking of bonds in molecular and/or reactive system. It has been amply demonstrated elsewhere, e.g., [7-22], that many classical problems in chemistry can be approached anew using this novel IT perspective.

The concepts and techniques of IT have been successfully used to explore the chemical properties of molecules and their fragments, to examine the bonding patterns in both molecular and reactive systems. For example, the displacements in the information distribution in molecules, relative to the *promolecular* reference consisting of the non-bonded constituent atoms, has been investigated [7-10] and the least-biased partition of the molecular electron distributions into the subsystem contributions, e.g., densities of bonded *Atoms-in-Molecules* (AIM), have been investigated [7,11-17]. This IT approach has been shown to lead to the *"stockholder"* molecular fragments of Hirshfeld [23], the density pieces of which have been derived from alternative global and local variational principles of IT.

Spatial localization of specific bonds, not to mention some qualitative questions about the very existence of some chemical bonds, e.g., between the bridgehead carbon atoms in propellanes, presents another challenging problem for this novel IT treatment of molecular systems. Another important aspect of the molecular electronic structure deals with the shell-structure and electron localization/delocalization in atoms and molecules. The non-additive Fisher information in *Atomic Orbital* (AO) resolution has been recently used as the *Contra-Gradience* (CG) criterion for localizing the bonding regions in molecules [19,24,25], while the related information density in the *Molecular Orbital* (MO) resolution has been shown [7,26] to determine the vital ingredient of the *Electron-Localization Function* (ELF) [27].

The *Communication Theory of the Chemical Bond* (CTCB) has been developed using the basic entropy/information descriptors of the *molecular information* (communication) *channels* defined in the AIM, orbital and local resolution levels of the electron probability distributions [7,8,28-32]. The same bond descriptors have been used to provide the information-scattering perspective on some intermediate stages in the electron redistribution processes [33a], including the atom promotion via the orbital hybridization [33b], and the communication theory for excited electron configurations has been developed [8b,34]. Moreover, the elements of a phenomenological description of equilibria in molecular subsystems, which resembles the ordinary thermodynamical approach, have been established [7,35].

To summarize, the entropic probes of the molecular electronic structure have provided novel attractive tools for describing the chemical bond phenomenona in information terms. It is the main purpose of this book to introduce the key IT concepts and techniques, as well as to examine some recent developments in alternative local entropy/information probes of the molecular electronic structure and in the orbital formulation of CTCB, called the *Orbital Communication Theory* (OCT). The information roots of quantum mechanics will be stressed and the importance of the non-additive effects in the chemical bond phenomena will be emphasized. The Schrödinger equations of quantum mechanics, which determine the equilibrium distribution of electrons and its time dependence for the fixed external potential due to the system fixed nuclei, in the familiar Born-Oppenheimer (BO) approximation, will be shown to result from the constrained-extremum principles of the Fisher information related to the system average kinetic energy [2,24,36-38].

The use of alternative IT probes of the electronic structure and chemical bonds in molecules will be illustrated. The electron redistribution accompanying the bond formation, from the (molecularly placed) free atoms $\{X^0\}$ of the "promolecule" consisting of the ground-state atomic densities $\{\rho_X{}^0\}$ to the bonded AIM with densities $\{\rho_X\}$, is marked by the difference $\Delta\rho = \rho - \rho^0$ between the molecular ($\rho = \Sigma_X\rho_X$) and promolecular ($\rho^0 = \Sigma_X\rho_X{}^0$) electron densities, which mark the process final and initial stages, respectively. The entropy/information content of these distributions provides a basis for a novel IT perspective on the chemical bond origins. The densities of displacements in the Shannon entropy and missing information (cross entropy) relative to the promolecular reference, will be used as sensitive diagnostic tools for detecting the chemical bonds, and to monitor the promotion/hybridization changes the bonded atoms undergo in their molecular environment. When applied to the classical problem of density partitioning into the AIM pieces the IT approach gives the familiar "stockholder" division scheme of Hirshfeld. The non-additive Fisher information density in the MO and AO resolutions generate the ELF and CG probes for localizing electrons and bonding regions in molecules, respectively. Representative examples of such an exploration of molecular ground-states will be presented.

The OCT of the chemical bond uses the standard entropy/information descriptors of the Shannon theory of communication to characterize the scattering of the AO electron probabilities throughout the network of chemical bonds generated by the system occupied MO. The molecule is thus treated as an information system which propagates the "signals" of electron allocations to these basis functions of the molecular SCF LCAO MO calculations, from the channel AO "inputs" to its AO "outputs". The underlying conditional probabilities of this "noisy" transmission process are generated from the bond-projected superposition principle of quantum mechanics. They are proportional to the squares of the corresponding elements of the *first*-order density matrix in the AO representation [8,20,32], thus being related to Wiberg's quadratic index of the chemical bond multiplicity [39].

Such information propagation in molecules exhibits the communication "noise", due to the electron delocalization via the system chemical bonds, which effectively lowers the information content in the *output*-signal distribution, compared to that contained in probabilities determining the channel *input*-signal, molecular, promolecular, or specifically designed to probe the properties of interest. The orbital information systems are now being used to generate the entropic measures of the chemical bond multiplicity and their covalent/ionic composition for both the molecule as a whole and its diatomic fragments. The *conditional-entropy*, which measures the channel average communication noise due to electron delocalization, reflects the IT-*covalency* in the molecule. The complementary descriptor of the the electron localization effects in the molecular information network, called the *mutual-information* in the channel inputs and outputs, which reflects the information-flow in the AO communication system, measures the system IT-*ionic* component. The illustrative examples of applying these novel IT tools for exploring the electronic and bonding structures of representative molecules will be reported and discussed. Other applications of the IT concepts to selected classical issues of quantum chemistry will be also briefly summarized.

ROMAN F. NALEWAJSKI Cracow, 2009

Chapter 1

INTRODUCTION TO INFORMATION THEORY

1.1. ENTROPY AND INFORMATION

The familiar *Shannon* (S) entropy $S[p]$ [3] of the (normalized) spatial probability distribution $p(\boldsymbol{r})$,

$$S[p] \equiv I^S[p] = -\int p(\boldsymbol{r}) \log p(\boldsymbol{r})\, d\boldsymbol{r}, \qquad \int p(\boldsymbol{r})\, d\boldsymbol{r} = 1, \tag{1}$$

where the (definite) integration is over the whole range of the random (position) variable $\boldsymbol{r}$, provides a measure of the average indeterminacy in $p(\boldsymbol{r})$ for the locality events $\{\boldsymbol{r}\}$. As also indicated above, this quantity also signifies the average amount of information $I^S[p]$ obtained when this spatial uncertainty is removed by an appropriate localization measurement (experiment). Here the logarithm is taken to an arbitrary but fixed base: when taken to base 2, $\log = \log_2$, the information is measured in *bits*, while selecting $\log = \ln$ expresses the information content in *nats*: 1 nat = 1.44 bits.

The *Fisher* (F) information [1,2] measure for locality, called the *intrinsic accuracy*, historically predates the Shannon entropy by about 25 years, being proposed in about the same time when the final form of the quantum mechanics was shaped. It emerges as an expected error in a "smart" measurement, in the context of the efficient estimators of a parameter. Fisher was the first to suggest that data samples in an experiment together with a given parametric distribution model contain the statistical information about the parameter(s).

Let $p(\boldsymbol{r}|\theta)$ be the probability distribution function depending upon the parameter θ. The Fisher measure of the information contained in this probability density is then defined as follows:

$$I(\theta) \equiv \int p(\boldsymbol{r}|\theta)\left(\frac{\partial \ln p(\boldsymbol{r}|\theta)}{\partial\theta}\right)^2 d\boldsymbol{r} = \int \frac{[p'(\boldsymbol{r}|\theta)]^2}{p(\boldsymbol{r}|\theta)} d\boldsymbol{r}, \qquad (2)$$

The *intrinsic accuracy* is a special case of this parametric measure, for the locality parameter $\boldsymbol{\theta}$, when $p(\boldsymbol{r}|\ \boldsymbol{\theta}) = p(\boldsymbol{r} + \boldsymbol{\theta}) = p(\boldsymbol{r}')$. In this case the parametric Fisher information provides the information about the probability distribution itself:

$$\frac{\partial p(\boldsymbol{r}|\boldsymbol{\theta})}{\partial \boldsymbol{\theta}} = \frac{\partial \boldsymbol{r}'}{\partial \boldsymbol{\theta}}\frac{\partial p(\boldsymbol{r}')}{\partial \boldsymbol{r}'} = \frac{\partial p(\boldsymbol{r}')}{\partial \boldsymbol{r}'} = \nabla p(\boldsymbol{r}'). \qquad (3)$$

Hence, for a *single*-component probability distribution

$$I(\boldsymbol{\theta}) = I[p] = \int p(\boldsymbol{r}')\,[\nabla \ln p(\boldsymbol{r}')]^2\, d\boldsymbol{r}' = \int [\nabla p(\boldsymbol{r})]^2/p(\boldsymbol{r})\, d\boldsymbol{r} \equiv I^F[p]. \qquad (4)$$

This information functional, reminiscent of von Weizsäcker's [36] inhomogeneity correction to the electronic kinetic energy in the Thomas-Fermi theory, characterizes the compactness ("order") of the probability density $p(\boldsymbol{r})$. For example, the Fisher information in the *normal distribution* measures the inverse of its *variance*, called the *invariance*, while the complementary Shannon entropy is proportional to the logarithm of variance, thus monotonically increasing with the spread of the Gaussian distribution.

The Shannon entropy and the Fisher information for locality thus describe the complementary facets of the probability density: the former reflects distribution's "spread" (a measure of uncertainty, "disorder"), while the latter measures its "narrowness" ("order"). The analytical properties of the Shannon and Fisher information functionals are quite different [2]. When extremized through variation of the probability distribution the Shannon entropy gives an exponential solution, while the Fisher information gives rise to differential equations, and hence to multiple solutions specified by the appropriate boundary conditions.

The form of the above intrinsic-accuracy functional can be simplified by expressing it as a functional of the associated classical (real) amplitude $A(\boldsymbol{r}) = \sqrt{p(\boldsymbol{r})}$ of the (non-negative) probability distribution $p(\boldsymbol{r}) = A^2(\boldsymbol{r})$:

$$I[p] = 4 \int [\nabla A(\boldsymbol{r})]^2\, d\boldsymbol{r} \equiv I[A]. \qquad (5)$$

It is naturally generalized into the realm of complex probability amplitudes one encounters in quantum mechanics, i.e., the system wave-functions [2,24]. For the simplest case of the spinless *one*-particle system, when $A(\boldsymbol{r}) = \psi(\boldsymbol{r})$ and $p(\boldsymbol{r}) = \psi^*(\boldsymbol{r})\,\psi(\boldsymbol{r}) = |\psi(\boldsymbol{r})|^2$,

$$I[\psi] = 4 \int |\nabla \psi(\boldsymbol{r})|^2\, d\boldsymbol{r} = 4 \int \nabla \psi^*(\boldsymbol{r})\cdot\nabla \psi(\boldsymbol{r})\, d\boldsymbol{r} \equiv \int f(\boldsymbol{r})\, d\boldsymbol{r}. \tag{6}$$

Therefore, the Fisher information $I[A]$ or $I[\psi]$ measures the gradient content in the amplitude of the probability-density. Its extension to the *multi*-component (vector) probabilities $\boldsymbol{p}(\boldsymbol{r}) \equiv \{p_n(\boldsymbol{r})\}$ expressed in terms of the associated probability amplitudes $\boldsymbol{A}(\boldsymbol{r}) \equiv \{A_n(\boldsymbol{r}) \equiv p_n(\boldsymbol{r})^{1/2}\}$ or $\boldsymbol{\psi} \equiv \{\psi_n(\boldsymbol{r})\}$ with $p_n(\boldsymbol{r}) = |\psi_n(\boldsymbol{r})|^2$ reads [2]:

$$I[\boldsymbol{p}] = \sum_n I[p_n] = 4\sum_n \int [\nabla A_n(\boldsymbol{r})]^2 d\boldsymbol{r} = \sum_n I[A_n] \equiv I[\boldsymbol{A}]$$

or

$$I[\boldsymbol{p}] = 4\sum_n \int |\nabla \psi_n(\boldsymbol{r})|^2 d\boldsymbol{r} = \sum_n I[\psi_n] \equiv I[\boldsymbol{\psi}]\cdot \tag{7}$$

1.2. Relative Measures of Information Content

An important generalization of Shannon's entropy, called the *relative* (*cross*) *entropy*, also known as the *entropy deficiency*, *missing information* or the *directed divergence*, has been proposed by *Kullback* and *Leibler* (KL) [4,5]. It measures the *information "distance"* between the two (normalized) probability distributions for the same set of elementary events. For example, in the discrete probability scheme $\mathbf{A} = [\boldsymbol{a}, \boldsymbol{P}(\boldsymbol{a})]$, involving events, identified by the admissible values $\boldsymbol{a} = \{a_i\}$ of the discrete random variable a, and their probabilities $\boldsymbol{P}(\boldsymbol{a}) = \{P(a_i) = p_i\} \equiv \boldsymbol{p}$, this discrimination information in $\boldsymbol{p}$ with respect to the *reference* distribution $\boldsymbol{p}^0 = \{P(a_i^0) = p_i^0\}$ reads:

$$\Delta S(\boldsymbol{p}|\boldsymbol{p}^0) \equiv I^{KL}(\boldsymbol{p}|\boldsymbol{p}^0) = \Sigma_i\, p_i \log(p_i/p_i^0) \geq 0, \tag{8}$$

In the continuous distribution case, e.g., $\mathbf{A} = \{\boldsymbol{r}, p(\boldsymbol{r})\}$, this directed-divergence measure of the entropy deficiency in the probability density $p(\boldsymbol{r})$, relative to the prior distribution $p^0(\boldsymbol{r})$, is defined by the related functional:

$$\Delta S[p|p^0] \equiv I^{KL}[p|p^0] = \int p(\boldsymbol{r}) \log[p(\boldsymbol{r})/p^0(\boldsymbol{r})]\, d\boldsymbol{r} \geq 0. \tag{9}$$

For individual events the logarithm of probability ratio, $I_i = \log(p_i/p_i^0)$, $I(\boldsymbol{r}) = \log[p(\boldsymbol{r})/p^0(\boldsymbol{r})]$ or for *one*-dimensional continuous distribution $I(x) = \log[p(x)/p^0(x)]$, called the *surprisal*, provides a measure of the information in the current distribution relative to the reference distribution. The equality in the two preceding equations takes place only when the surprisal vanishes for all events, i.e., when the compared probability schemes are identical.

The non-negative character of the entropy deficiency in $p(x)$ relative to $p^0(x)$,

$$\Delta S[p \,|\, p^0] = \int p(x) \log[p(x)/p^0(x)]\, dx \equiv \int p(x)\, I(x)\, dx, \tag{10}$$

directly follows from the observation that the line $y = z - 1$ lies above the curve $y = \log z$, with two functions having equal (zero) value only for $z = 1$. Taking $z(x) = p^0(x)/p(x)$ and using the condition of the probability normalization then gives:

$$\Delta S[p \,|\, p^0] = -\int p(x) \log z(x)\, dx$$
$$\geq \int p(x)\, [z(x) - 1]\, dx = \int [p^0(x) - p(x)]\, dx = 0. \tag{11}$$

Therefore, the entropy deficiency provides a measure of an information resemblance between the two compared probability schemes. The more the two probability distributions differ from one another, the larger the information distance. As we have already remarked before, the Shannon entropy measures a "*disorder*" (uncertainty, indeterminacy, "smoothness") of the probability distribution. On a finite interval, the distribution possessing the highest entropy is the uniform distribution, and any deviation from uniformity indicates the "perturbing" presence of *"order"*. The Kullback-Leibler measure, i.e., the *referenced* Shannon's entropy, provides a similar description, but in reference to some prior distribution.

Notice, however, that the surprisal itself becomes negative, when the current probability is lower than the reference value. In the continuous case the directed-divergence functional is not symmetrical with respect to the two probability distributions involved and exhibits negative values of the integrand. To avoid this limitation Kullback (K) [5] has proposed an alternative measure, called the *divergence*, defined by the symmetrized combination of the two admissible entropy deficiencies (directed-divergencies), which gives rise to a *non*-negative integrand:

$$\Delta S(\boldsymbol{p}, \boldsymbol{p}^0) \equiv I^{\mathrm{K}}(\boldsymbol{p}, \boldsymbol{p}^0) = \Delta S(\boldsymbol{p}|\boldsymbol{p}^0) + \Delta S(\boldsymbol{p}^0|\boldsymbol{p})$$

$$= \sum_i (p_i - p_i^0)\, I[p_i/p_i^0] \equiv \sum_i \Delta p_i\, I_i, \qquad \Delta p_i\, I_i \geq 0,$$

$$\Delta S[p, p^0] \equiv I^{\mathrm{K}}[p, p^0] = \Delta S[p|p^0] + \Delta S[p^0|p]$$

$$= \int [p(\boldsymbol{r}) - p^0(\boldsymbol{r})]\, I(\boldsymbol{r})\, d\boldsymbol{r} \equiv \int \Delta p(\boldsymbol{r})\, I(\boldsymbol{r})\, d\boldsymbol{r}, \qquad \Delta p(\boldsymbol{r})\, I(\boldsymbol{r}) \geq 0. \qquad (12)$$

1.3. Dependent Probability Distributions

For two mutually dependent (discrete) probability distributions,

$$\boldsymbol{P}(\boldsymbol{a}) = \{P(a_i) = p_i\} \equiv \boldsymbol{p} \qquad \text{and} \qquad \boldsymbol{P}(\boldsymbol{b}) = \{P(b_j) = q_j\} \equiv \boldsymbol{q}, \qquad (13)$$

determining the associated probability schemes $\mathbf{A} = [\boldsymbol{a}, \boldsymbol{P}(\boldsymbol{a})]$ and $\mathbf{B} = [\boldsymbol{b}, \boldsymbol{P}(\boldsymbol{b})]$, respectively, we decompose the joint-probabilities $\mathbf{P}(\boldsymbol{a}\otimes\boldsymbol{b}) = \{P(a_ib_j) = \pi_{i,j}\} \equiv \boldsymbol{\pi}$ of the simultaneous events $\boldsymbol{a}\wedge\boldsymbol{b} = \{a_i\wedge b_j \equiv a_ib_j\} \equiv \boldsymbol{a}\otimes\boldsymbol{b}$ as products $\{\pi_{i,j} = p_i\, P(j|i)\}$ of the *marginal* probability $p_i = P(a_i)$ of ith event in the $\boldsymbol{a}$ set of events, and the *conditional* probability $P(j|i) = P(a_ib_j)/P(a_i)$ of the event b_j of the other set $\boldsymbol{b}$ of outcomes, given that the event a_i has already occurred. The relevant normalization conditions for the joint probabilities $\boldsymbol{\pi}$ and the conditional probabilities $\mathbf{P}(\boldsymbol{b}|\boldsymbol{a}) = \{P(j|i)\}$ then read:

$$\sum_j \pi_{i,j} = p_i, \quad \sum_i \pi_{i,j} = q_j, \quad \sum_i\sum_j \pi_{i,j} = 1,$$

$$\sum_j P(j|i) = 1, \qquad i, j = 1, 2, \ldots, n. \qquad (14)$$

In this scenario the Shannon entropy of the product distribution $\boldsymbol{\pi} = \{\pi_{i,j}\}$,

$$S(\boldsymbol{\pi}) = -\sum_i\sum_j \pi_{i,j} \log\pi_{i,j} = -\sum_i\sum_j p_i\, P(j|i)\, [\log p_i + \log P(j|i)]$$

$$= -\sum_i [\sum_j P(j|i)]\, p_i \log p_i - \sum_i p_i\, [\sum_j P(j|i) \log P(j|i)]$$

$$\equiv S(\boldsymbol{p}) + \sum_i p_i\, S(\boldsymbol{q}|i) \equiv S(\boldsymbol{p}) + S(\boldsymbol{q}|\boldsymbol{p}), \qquad (15)$$

is seen to be given by the sum of the average entropy in the marginal probability distribution, $S(\boldsymbol{p})$, and the average *conditional entropy* $S(\boldsymbol{q}|\boldsymbol{p})$ in $\boldsymbol{q}$ given $\boldsymbol{p}$. The latter appears as the weighted average of the conditional entropies $\{S(\boldsymbol{q}|i)\}$ due to each input:

$$S(\boldsymbol{q}|\boldsymbol{p}) = \sum_i p_i\,[-\sum_j P(j|i)\,\log P(j|i)] \equiv \sum_i p_i\,S(\boldsymbol{q}|i). \tag{16}$$

It represents the extra amount of uncertainty about the occurrence of events $\boldsymbol{b}$, given that the events $\boldsymbol{a}$ are known to have occurred. In other words: the amount of information obtained as a result of simultaneously observing the events $\boldsymbol{a}$ and $\boldsymbol{b}$ of the two discrete probability distributions $\boldsymbol{p} = \boldsymbol{P}(\boldsymbol{a})$ and $\boldsymbol{q} = \boldsymbol{P}(\boldsymbol{b})$, respectively, equals to the amount of information observed in one set, say $\boldsymbol{a}$, supplemented by the extra information provided by the occurrence of events in the other set $\boldsymbol{b}$, when $\boldsymbol{a}$ are known to have occurred already. This is qualitatively illustrated in Scheme 1 (see also the upper part of Scheme 4).

Clearly, by using the other probability distribution $\boldsymbol{q} = \{P(b_i)\}$ as the marginal factor of the joint probabilities one arrives at the alternative expression for $S(\boldsymbol{\pi})$:

$$S(\boldsymbol{\pi}) = S(\boldsymbol{q}) + \sum_j q_j\,S(\boldsymbol{p}|j) \equiv S(\boldsymbol{q}) + S(\boldsymbol{p}|\boldsymbol{q}), \tag{17}$$

where $S(\boldsymbol{p}|\boldsymbol{q}) = \sum_j q_j\,S(\boldsymbol{p}|j)$ and the conditional probabilities $P(\boldsymbol{a}|\boldsymbol{b}) = \{P(i|j)\}$ are normalized to satisfy the conditions $\sum_i P(i|j) = 1, j = 1, 2, ..., m$. Here, the average conditional entropy $S(\boldsymbol{p}|\boldsymbol{q})$ represents the residual uncertainty about $\boldsymbol{a}$, when $\boldsymbol{b}$ are known to have been observed. Properties of the conditional entropy are similar to those of the information entropy itself, since the conditional probability is a probability measure.

The common amount of information in two events a_i and b_j, $I(i{:}j)$, measuring the information about a_i provided by the occurrence of b_j, or equivalently - the information about b_j provided by the occurrence of a_i, is called the *mutual information* in two events:

$$I(i{:}j) = \log[P(a_i b_j)/P(a_i)P(b_j)] = \log[\pi_{i,j}/(p_i\,q_j)] \equiv \log[\pi_{i,j}/\pi_{i,j}^{ind.}]$$

$$\equiv \log[P(i|j)/p_i] \equiv \log[P(j|i)/q_j] = I(j{:}i); \tag{18}$$

here $\pi_{i,j}^{ind.} = p_i q_j = P^{ind.}(a_i \wedge b_j)$ stands for the joint probability of two independent events a_i and b_j. The mutual information in two events may take on any real value, positive, negative, or zero. It vanishes, when both events are independent, i.e., when the occurrence of one event does not influence (or condition) the probability of the occurrence of the other event, and it is negative when the occurrence of one event makes a nonoccurrence of the other event more likely.

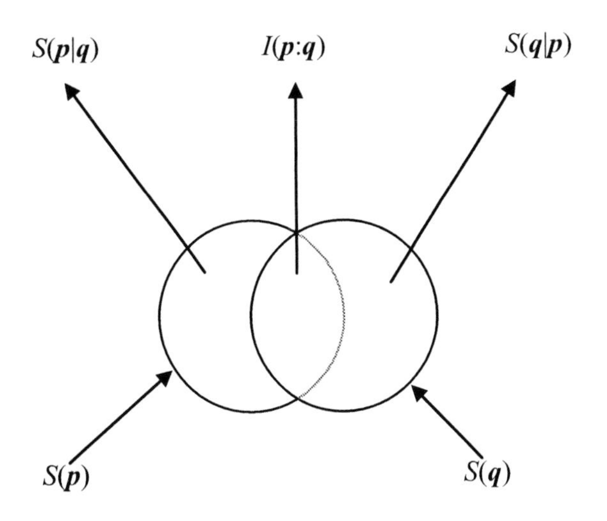

Scheme 1. A qualitative diagram of the conditional entropy and mutual information quantities for two dependent probability distributions $\boldsymbol{p}$ and $\boldsymbol{q}$. Two circles enclose the areas representing the entropies $S(\boldsymbol{p})$ and $S(\boldsymbol{q})$ of the two *separate* distributions. The common (overlap) area of the two circles corresponds to the *mutual information* $I(\boldsymbol{p}:\boldsymbol{q})$ in two distributions. The remaining parts of two circles represent the corresponding *conditional entropies* $S(\boldsymbol{p}|\boldsymbol{q})$ and $S(\boldsymbol{q}|\boldsymbol{p})$, measuring the residual uncertainty about events in one set, when one has the full knowledge of the occurrence of the events in the other set of outcomes. The area enclosed by the envelope of the two overlapping circles then represents the entropy of the "product" (joint) distribution: $S(\boldsymbol{\pi}) = S(\mathbf{P}(\boldsymbol{a}\otimes\boldsymbol{b})) = S(\boldsymbol{p}) + S(\boldsymbol{q}) - I(\boldsymbol{p}:\boldsymbol{q}) = S(\boldsymbol{p}) + S(\boldsymbol{q}|\boldsymbol{p}) = S(\boldsymbol{q}) + S(\boldsymbol{p}|\boldsymbol{q})$

It also follows from the preceding equation that

$$I(i:j) = I(i) - I(i|j) = I(j) - I(j|i) = I(i) + I(j) - I(i\wedge j) \qquad \text{or}$$

$$I(i\wedge j) = I(i) + I(j) - I(i:j), \tag{19}$$

where the self-information of the joint event $I(i\wedge j) = -\log\pi_{i,j}$. Thus, the information in the joint occurrence of two events a_i and b_j is the information in the occurrence of a_i plus that originating from the occurrence of b_j minus the mutual information. Clearly, for independent events $I(i, j) = I(i) + I(j)$, since then $I(i:j) = 0$.

The mutual information of an event with itself defines its *self-information*: $I(i{:}i) \equiv I(i) = \log[P(i|i)/p_i] = -\log p_i$, since $P(i|i) = 1$. It vanishes when $p_i = 1$, i.e., when there is no uncertainty about the occurrence of a_i so that the occurrence of this event removes no uncertainty, hence conveys no information. This quantity provides a measure of the uncertainty about the occurrence of the single event, i.e., the information received when this event occurs. The Shannon entropy of Eq. (1) can be thus interpreted as the mean value of the self-informations in all individual events defining the probability distribution, e.g., $S(\boldsymbol{p}) = -\sum_i p_i \log p_i \equiv \sum_i p_i\, I(i)$. One then defines the *average* mutual information in two probability distributions, $I(\boldsymbol{p}{:}\boldsymbol{q})$, as the $\boldsymbol{\pi}$-weighted mean value of the mutual information quantities for the individual joint events:

$$I(\boldsymbol{p}{:}\boldsymbol{q}) = \sum_i \sum_j \pi_{i,j}\, I(i{:}j) = \sum_i \sum_j \pi_{i,j} \log[\pi_{i,j} / \pi_{i,j}^{ind.}]$$

$$= S(\boldsymbol{p}) + S(\boldsymbol{q}) - S(\boldsymbol{\pi}) = S(\boldsymbol{p}) - S(\boldsymbol{p}|\boldsymbol{q}) = S(\boldsymbol{q}) - S(\boldsymbol{q}|\boldsymbol{p}) \geq 0. \qquad (20)$$

where the equality holds only for the independent distributions, when $\pi_{i,j}^{ind.} = \pi_{i,j}$. These average entropy/information relations are also illustrated in Scheme 1.

The non-negative character of this average information measure can again be demonstrated using the inequality $\log z \leq z - 1$: for $z = \pi_{i,j}^{ind.}/\pi_{i,j}$ the normalization conditions of $\boldsymbol{\pi} = \{\pi_{i,j}\}$ and $\boldsymbol{\pi}^{ind.} = \{\pi_{i,j}^{ind.}\}$ then give:

$$I(\boldsymbol{p}{:}\boldsymbol{q}) = \sum_i \sum_j \pi_{i,j} \log(\pi_{i,j}/\pi_{i,j}^{ind.}) \geq \sum_i \sum_j (\pi_{i,j} - \pi_{i,j}^{ind.}) = 0. \qquad (21)$$

Indeed, as directly seen in Scheme 1, the amount of uncertainty in $\boldsymbol{q}$ can only decrease, when $\boldsymbol{p}$ has been known beforehand, $S(\boldsymbol{q}) \geq S(\boldsymbol{q}|\boldsymbol{p}) = S(\boldsymbol{q}) - I(\boldsymbol{p}{:}\boldsymbol{q})$, with equality being observed only when the two sets of events are independent (non-overlapping).

It should be also observed that the average mutual information is an example of the entropy deficiency functional, which measures the missing information between the joint probabilities $\mathbf{P}(\boldsymbol{a}\otimes\boldsymbol{b}) = \boldsymbol{\pi}$ of the *dependent* events $\boldsymbol{a}$ and $\boldsymbol{b}$, and the joint probabilities $\mathbf{P}^{ind.}(\boldsymbol{a}\otimes\boldsymbol{b}) = \boldsymbol{\pi}^{ind.} \equiv \boldsymbol{p}\otimes\boldsymbol{q}$ for independent events: $I(\boldsymbol{p}{:}\boldsymbol{q}) = \Delta S(\boldsymbol{\pi}|\boldsymbol{\pi}^{ind.})$. The average mutual information thus measures a degree of a dependence between events defining the two probability schemes. A similar information-distance interpretation can be given to the conditional entropy: $S(\boldsymbol{p}|\boldsymbol{q}) = S(\boldsymbol{p}) - \Delta S(\boldsymbol{\pi}|\boldsymbol{\pi}^{ind.})$.

1.4. Communication Systems and Their Entropy/Information Descriptors

We continue this short overview with the key entropy/information quantities describing a transmission of signals in communication systems [3,6]. The basic elements of such a "device" are shown in Scheme 2. The *input*-signal emitted from n "inputs" $\boldsymbol{a} = (a_1, a_2, \ldots, a_n)$ of the channel *source* (**A**) is characterized by the input probability distribution $\boldsymbol{P(a)} = \boldsymbol{p} = (p_1, p_2, \ldots, p_n) \equiv \boldsymbol{P}(\mathbf{A})$. It can be received at m "outputs" $\boldsymbol{b} = (b_1, b_2, \ldots, b_m)$ of the system *receiver* (**B**). The distribution of the *output*-signal among the detection "events" $\boldsymbol{b}$ gives rise to the *output* probability distribution $\boldsymbol{P(b)} = \boldsymbol{q} = (q_1, q_2, \ldots, q_m) \equiv \boldsymbol{P}(\mathbf{B})$.

The transmission of signals is randomly disturbed within the communication system, thus exhibiting a typical communication *noise*. Indeed, the signal sent at the given input can be in general received with a non-zero probability at several outputs. This feature of communication systems is reflected by the conditional probabilities of the *outputs-given-inputs*, $\mathbf{P(B|A)} = \{P(b_j|a_i) = P(a_i \wedge b_j)/P(a_i) \equiv P(j|i)\}$, or the conditional probabilities of the *inputs-given-outputs*, $\mathbf{P(A|B)} = \{P(a_i|b_j) = P(a_i \wedge b_j)/P(b_j) \equiv P(i|j)\}$, where $P(a_i \wedge b_j) \equiv \pi_{i,j}$ stands for the probability of the joint occurrence of the specified pair of the (*output*$\wedge$*input*) events.

The Shannon entropy of the input (source) probabilities $\boldsymbol{p}$, $H(\mathbf{A}) \equiv S(\boldsymbol{p})$, determines the channel *a priori* entropy. The average *conditional entropy* $H(\mathbf{B|A}) \equiv S(\boldsymbol{q}|\boldsymbol{p})$, of the outputs given inputs, is determined by the scattering probabilities $\mathbf{P(B|A)} = \{P(b_j|a_i) \equiv P(j|i)\} \equiv \mathbf{P}(\boldsymbol{b}|\boldsymbol{a})$. It measures the average noise in the *"forward"* transmission of signals, from $\boldsymbol{a}$ to $\boldsymbol{b}$. The so called *a posteriori* entropy of the input, given output, $H(\mathbf{A|B}) \equiv S(\boldsymbol{p}|\boldsymbol{q})$, is similarly defined by the conditional probabilities of the *"reverse"* probability scattering, from $\boldsymbol{b}$ to $\boldsymbol{a}$: $\mathbf{P(A|B)} = \{P(a_i|b_j) = P(i|j)\} \equiv \mathbf{P}(\boldsymbol{a}|\boldsymbol{b})$. It reflects the residual indeterminacy about input signal, when the output signal has already been received.

The average conditional entropy $H(\mathbf{A|B}) \equiv S(\boldsymbol{p}|\boldsymbol{q})$ thus measures the indeterminacy of the source with respect to the receiver, while the conditional entropy $H(\mathbf{B|A}) \equiv S(\boldsymbol{q}|\boldsymbol{p})$ reflects the uncertainty of the receiver relative to the source. Hence, an observation of an output signal provides on average the amount of information given by the difference between the *a priori* and *a posteriori* uncertainties, $S(\boldsymbol{p}) - S(\boldsymbol{p}|\boldsymbol{q}) = I(\boldsymbol{p}:\boldsymbol{q}) \equiv I(\mathbf{A}:\mathbf{B})$, which defines the *mutual information* in the source and receiver. In other words, the mutual information measures the net amount of information transmitted through the communication channel, while the conditional information $S(\boldsymbol{p}|\boldsymbol{q})$ reflects a portion of $S(\boldsymbol{p})$

dissipated as the communication "noise" as a result of the input signal passing through the information channel. Accordingly, $S(\boldsymbol{q}|\boldsymbol{p})$ reflects the noise part of $S(\boldsymbol{q})$: $S(\boldsymbol{q}) = S(\boldsymbol{q}|\boldsymbol{p}) + I(\boldsymbol{p}:\boldsymbol{q})$.

Input (Source): **A** Communication network: $\mathbf{P}(\mathbf{B}|\mathbf{A})$ *Output* (Receiver): **B**

$$
\begin{array}{ccccc}
 & a_1 & & b_1 & \\
 & a_2 & & b_2 & \\
\cdots & \cdots & \cdots & \cdots & \cdots \\
p_i \rightarrow & a_i & \longrightarrow P(b_j|a_i) \equiv P(j|i) \longrightarrow & b_j & \rightarrow q_j \\
\cdots & \cdots & \cdots & \cdots & \cdots \\
 & a_n & & b_m &
\end{array}
$$

Scheme 2. A schematic diagram of the communication system, characterized by the probability vectors $\boldsymbol{P}(\boldsymbol{a}) = \{P(a_i)\} = \boldsymbol{p} = (p_1, \ldots, p_n) \equiv \boldsymbol{P}(\mathbf{A})$, of the channel *"input"* events $\boldsymbol{a} = (a_1, \ldots, a_n)$ in the system *source* **A**, and $\boldsymbol{P}(\boldsymbol{b}) = \{P(b_j)\} = \boldsymbol{q} = (q_1, \ldots, q_m) \equiv \boldsymbol{P}(\mathbf{B})$, of the *"output"* events $\boldsymbol{b} = (b_1, \ldots, b_m)$ in the system *receiver* **B**. The transmission of signals via a network of communication channels is described by the $(n{\times}m)$-matrix of the conditional probabilities $\mathbf{P}(\mathbf{B}|\mathbf{A}) = \{P(b_j|a_i) \equiv P(j|i)\} \equiv \boldsymbol{P}(\boldsymbol{b}|\boldsymbol{a})$, of observing different "outputs" (*columns*, $j = 1, 2, \ldots, m$), given the specified "inputs" (*rows*, $i = 1, 2, \ldots, n$). For clarity, only a single "forward" scattering $a_i \rightarrow b_j$ of the whole channel network is shown in the diagram

Consider, for example, the *Symmetric Binary Channel* (SBC) of Scheme 3, consisting of two inputs and two outputs with the input probabilities $\boldsymbol{p} = (x, 1 - x)$ and the symmetric conditional probability matrix

$$
\mathbf{P}(\mathbf{B}|\mathbf{A}) = \begin{bmatrix} 1-\omega & \omega \\ \omega & 1-\omega \end{bmatrix}. \tag{22}
$$

Its input entropy is determined by the *Binary Entropy Function* (BEF) $H(x)$ of Figure 1,

$$
H(\mathbf{A}) = -\,x\log x - (1 - x)\log(1 - x) \equiv H(x), \tag{23}
$$

which defines the channel *a priori* entropy. The system output entropy is also generated by BEF, $H(\mathbf{B}) = H(z(x, \omega))$, where $z(x, \omega) \equiv q_2 = x\omega + (1-x)(1-\omega)$ (see Scheme 3).

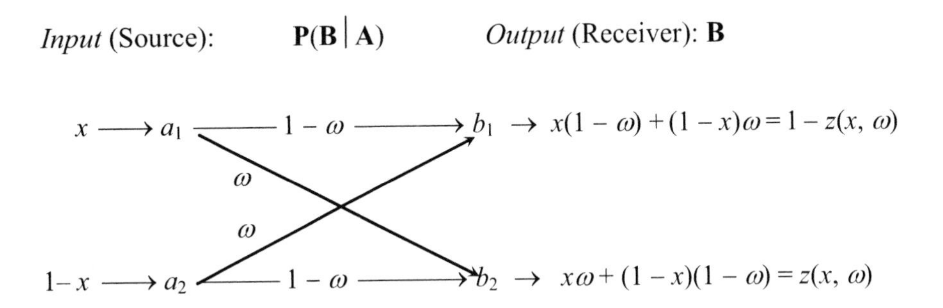

Scheme 3. The symmetric binary channel (SBC)

One also finds that the channel conditional entropy $H(\mathbf{B}|\mathbf{A}) = S(\boldsymbol{q}|\boldsymbol{p}) = H(\omega)$ measures its average communication noise in the forward transmission of signals. Hence, the mutual information between the system inputs and outputs,

$$I(\mathbf{A}:\mathbf{B}) = S(\mathbf{B}) - H(\mathbf{B}|\mathbf{A}) = H[z(x, \omega)] - H(\omega), \tag{24}$$

reflects the net information flow in SBC. These relations are illustrated in Figure 1. It should be observed that z always lies between ω and $1-\omega$, so that $H(z) = H(1 - z) \geq H(\omega) = H(1 - \omega)$. This demonstrates a *non*-negative character of the mutual information, represented by the overlap areas between the two entropy circles in qualitative diagrams of Schemes 1 and 4.

The amount of information $I(\mathbf{A}:\mathbf{B})$ flowing through SBC thus depends on both the conditional probability parameter ω, characterizing the communication system itself, and on the input probability parameter x, which determines the way the channel is used (or probed). For $x = 0$ (or $x = 1$) $H(z) = H(\omega)$ and thus $I(\mathbf{A}:\mathbf{B}) = 0$, i.e., there is no net flow of information from the source to the receiver whatever. For $x = ½$, when the two input signals are equally probable, one finds $H(z) = 1$ bit, thus giving rise to the maximum value of the channel mutual information, determining the system transmission *capacity*

$$C(\omega) \equiv \max_{\mathbf{A}} I(\mathbf{A}{:}\mathbf{B}) = \max_x\{H[z(x, \omega)] - H(\omega)\} = 1 - H(\omega) \text{ (in bits)}. \quad (25)$$

Hence, for $\omega = ½$, the information capacity of SBC identically vanishes.

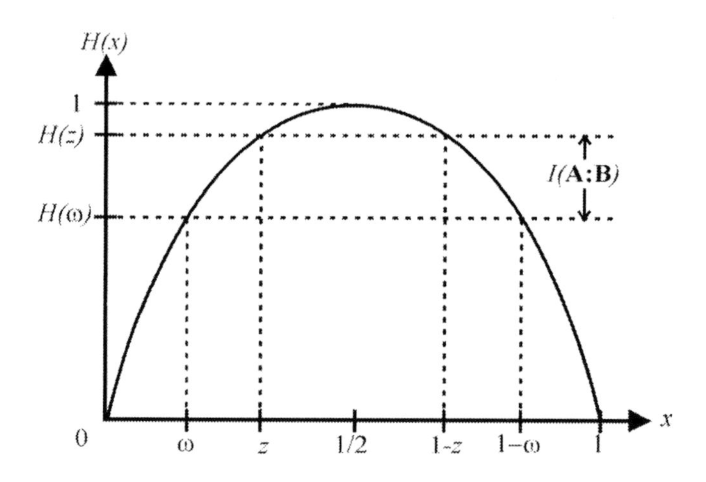

Figure 1. The binary entropy function (in bits), $H(x) = -x\log_2 x - (1-x)\log_2(1-x)$, and the geometric interpretation of the conditional entropy $H(\mathbf{B}|\mathbf{A}) = H(\omega)$ and the mutual information $I(\mathbf{A}{:}\mathbf{B}) = H(z) - H(\omega)$ in the SBC of Scheme 3

As we have already remarked in the Preface, in the molecular communication system [7], when $\boldsymbol{p}$ and $\boldsymbol{q}$ denote the channel input and output probabilities, the conditional entropy $S(\boldsymbol{q}|\boldsymbol{p})$ measures the entropy-covalency of all bonds in the molecule as a whole, while the mutual information relative to the reference ("promolecular") distribution $\boldsymbol{p}^0 = (½,½)$ of the SBC, $I(\boldsymbol{p}^0{:}\boldsymbol{q}) = C(\omega)$, reflects the channel IT-ionic component. It thus follows that the overall IT bond index of SBC,

$$N(\mathbf{A}^0; \mathbf{B}) = N(\boldsymbol{p}^0; \boldsymbol{q}) = S(\boldsymbol{q}|\boldsymbol{p}) + I(\boldsymbol{p}^0{:}\boldsymbol{q}) = 1 \text{ (bit)} \quad (26)$$

preserves the reference input entropy $H(\mathrm{A}^0) = S(p^0) = H(½) = 1$ bit.

1.5. SEVERAL PROBABILITY SCHEMES

The lower part of Scheme 4 shows a variety of the entropy/information descriptors of three dependent (overlapping) probability schemes [6]: $\mathbf{A} = [\boldsymbol{a}, \boldsymbol{P}(\boldsymbol{a})]$, $\mathbf{B} = [\boldsymbol{b}, \boldsymbol{P}(\boldsymbol{b})]$, and $\mathbf{C} = [\boldsymbol{c}, \boldsymbol{P}(\boldsymbol{c})]$, while the diagram upper part summarizes quantities characterizing the *two*-distribution case of Scheme 1. In what follows we denote by a, b and c, the representative events in these three probability schemes: $a \in \boldsymbol{a}$, $b \in \boldsymbol{b}$ and $c \in \boldsymbol{c}$. These entropy/information data now involve the conditional (relative) entropies with respect to a single or two probability schemes, the mutual information contained in two or three probability distributions, and the conditional mutual-information characteristics of three probability schemes.

The diagrams in the lower part of Scheme 4 demonstrate the *additivity* of the information quantities in the three-probability scenario, which may now involve two *output*-signals resulting from a single *input*-signal in the communication channel, or the repeated, two *input*-signals transmitted in the noisy information system. For example, the information about the input signal **A** obtained by observing two output signals of **B** and **C** is measured by the difference of the Shannon (*a priori*) entropy $H(\mathbf{A})$ and the conditional (double *a posteriori*) entropy in **A** given **B** and **C**, i.e., the indeterminacy of **A** relative to (**B** *and* **C**), which defines the *three*-scheme mutual-information quantity

$$
\begin{aligned}
I(\mathbf{A{:}BC}) \equiv I(\mathbf{A{:}B}\wedge\mathbf{C}) &= \sum_{a\in\boldsymbol{a}}\sum_{b\in\boldsymbol{b}}\sum_{c\in\boldsymbol{c}} P(a\wedge b\wedge c)\log\frac{P(a\wedge b\wedge c)}{p(a)P(b\wedge c)} \\
&= \sum_{a\in\boldsymbol{a}}\sum_{b\in\boldsymbol{b}}\sum_{c\in\boldsymbol{c}} P(a\wedge b\wedge c)\log\frac{P(a|b\wedge c)}{p(a)} \\
&= H(\mathbf{A}) - H(\mathbf{A}|\mathbf{BC}) \\
&= S(\boldsymbol{P}(\boldsymbol{a})) - S(\boldsymbol{P}(\boldsymbol{a})|\mathbf{P}(\boldsymbol{b}\wedge\boldsymbol{c})) \equiv I(\boldsymbol{P}(\boldsymbol{a}){:}\mathbf{P}(\boldsymbol{b}\wedge\boldsymbol{c})),
\end{aligned}
\qquad (27)
$$

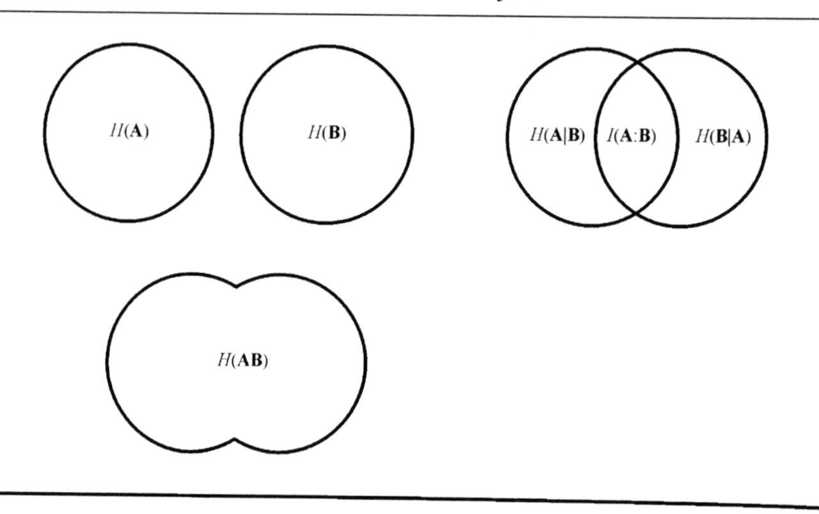

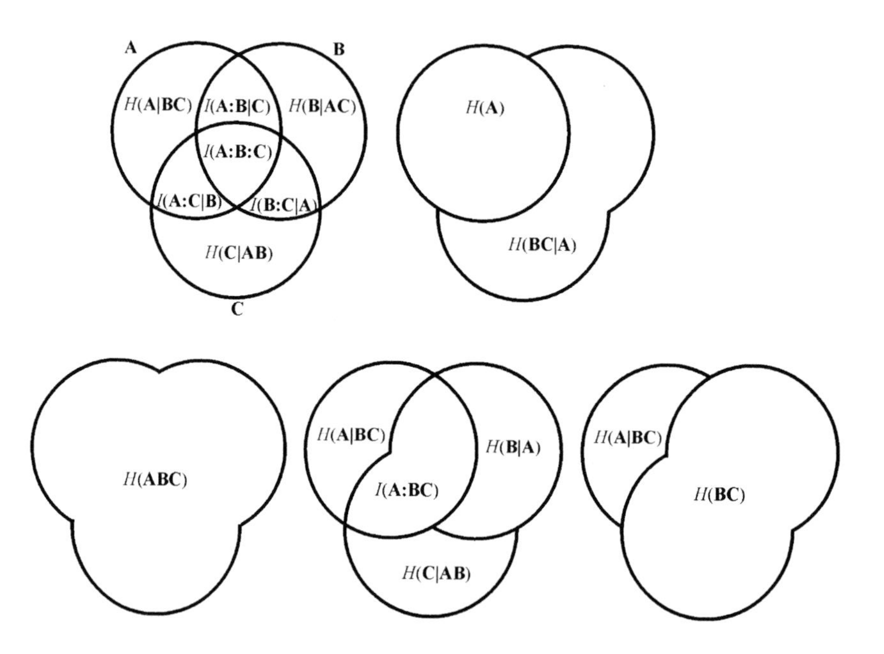

Scheme 4. A variety of the conditional-entropy and mutual-information descriptors of two (upper part) and three (lower part) dependent probability schemes [6]. Here, the individual circles $H(\mathbf{A}) = S(\boldsymbol{P}(\boldsymbol{a}))$, $H(\mathbf{B}) = S(\boldsymbol{P}(\boldsymbol{b}))$ and $H(\mathbf{C}) = S(\boldsymbol{P}(\boldsymbol{c}))$ represent the Shannon entropies of the separate probability distributions, $H(\mathbf{AB}) = S(\mathbf{P}(\boldsymbol{a}\otimes\boldsymbol{b})) = S(\mathbf{P}(\boldsymbol{a}\wedge\boldsymbol{b}))$, $H(\mathbf{B}|\mathbf{A}) = S(\boldsymbol{P}(\boldsymbol{b})|\boldsymbol{P}(\boldsymbol{a}))$, $I(\mathbf{A}:\mathbf{B}) = I(\boldsymbol{P}(\boldsymbol{a}):\boldsymbol{P}(\boldsymbol{b}))$, etc

where:

$$H(\mathbf{A}|\mathbf{BC}) = -\sum_{a\in\boldsymbol{a}}\sum_{b\in\boldsymbol{b}}\sum_{c\in\boldsymbol{c}} P(a\wedge b\wedge c)\log P(a|b\wedge c) \equiv S(\boldsymbol{P}(\boldsymbol{a})|\mathbf{P}(\boldsymbol{b}\wedge\boldsymbol{c})). \qquad (28)$$

The mutual information quantity of Eq. (27) can be alternatively expressed in terms of other entropic quantities shown in the qualitative diagrams of Scheme 4. For example, it is seen to measure the sum of elementary amounts of information in three or two probability schemes, or it can be expressed in terms of the relevant entropies:

$$I(\mathbf{A}:\mathbf{BC}) = I(\mathbf{A}:\mathbf{B}|\mathbf{C}) + I(\mathbf{A}:\mathbf{C}|\mathbf{B}) + I(\mathbf{A}:\mathbf{B}:\mathbf{C})$$

$$\equiv I(\boldsymbol{P}(\boldsymbol{a}):\boldsymbol{P}(\boldsymbol{b})|\boldsymbol{P}(\boldsymbol{c})) + I(\boldsymbol{P}(\boldsymbol{a}):\boldsymbol{P}(\boldsymbol{c})|\boldsymbol{P}(\boldsymbol{b})) + I(\boldsymbol{P}(\boldsymbol{a}):\boldsymbol{P}(\boldsymbol{b}):\boldsymbol{P}(\boldsymbol{c})), \qquad (29)$$

$$I(\mathbf{A}:\mathbf{BC}) = I(\mathbf{A}:\mathbf{B}) + I(\mathbf{A}:\mathbf{C}|\mathbf{B}) \equiv I(\boldsymbol{P}(\boldsymbol{a}):\boldsymbol{P}(\boldsymbol{b})) + I(\boldsymbol{P}(\boldsymbol{a}):\boldsymbol{P}(\boldsymbol{c})|\boldsymbol{P}(\boldsymbol{b})), \qquad (30)$$

$$I(\mathbf{A}:\mathbf{BC}) = H(\mathbf{AB}) - H(\mathbf{A}|\mathbf{BC}) - H(\mathbf{B}|\mathbf{A})$$

$$\equiv S(\mathbf{P}(\boldsymbol{a}\wedge\boldsymbol{b})) - S(\boldsymbol{P}(\boldsymbol{a})|\mathbf{P}(\boldsymbol{b}\wedge\boldsymbol{c})) - S(\boldsymbol{P}(\boldsymbol{b})|\boldsymbol{P}(\boldsymbol{a})). \qquad (31)$$

Here, the mutual information in two probability distributions, given the third distribution, reads:

$$I(\mathbf{A}:\mathbf{B}|\mathbf{C}) = H(\mathbf{A}|\mathbf{C}) - H(\mathbf{A}|\mathbf{BC})$$

$$= \sum_{a\in\boldsymbol{a}}\sum_{b\in\boldsymbol{b}}\sum_{c\in\boldsymbol{c}} P(a\wedge b\wedge c)\log\frac{P(a|b\wedge c)}{P(a|c)}$$

$$= \sum_{a\in\boldsymbol{a}}\sum_{b\in\boldsymbol{b}}\sum_{c\in\boldsymbol{c}} P(a\wedge b\wedge c)\log\frac{P(a\wedge b|c)}{P(a|c)P(b|c)}$$

$$= \sum_{a\in\boldsymbol{a}}\sum_{b\in\boldsymbol{b}}\sum_{c\in\boldsymbol{c}} P(a\wedge b\wedge c)\log\frac{P(a\wedge b\wedge c)p(c)}{P(a\wedge c)P(b\wedge c)}. \qquad (32)$$

The mutual information in three probability schemes is represented in Scheme 4 by the common area of three entropy circles and may assume negative values.

This quantity can be similarly expressed in terms of the elementary mutual informations or entropies:

$$I(\mathbf{A}:\mathbf{B}:\mathbf{C}) = I(\mathbf{A}:\mathbf{B}) - I(\mathbf{A}:\mathbf{B}|\mathbf{C}) = I(\mathbf{A}:\mathbf{C}) - I(\mathbf{A}:\mathbf{C}|\mathbf{B}) = I(\mathbf{B}:\mathbf{C}) - I(\mathbf{B}:\mathbf{C}|\mathbf{A})$$

$$= \sum_{a\in\boldsymbol{a}}\sum_{b\in\boldsymbol{b}}\sum_{c\in\boldsymbol{c}} P(a\wedge b\wedge c)\log\frac{P(a\wedge b)P(a\wedge c)P(b\wedge c)}{p(a)p(b)p(c)P(a\wedge b\wedge c)}$$

$$= H(\mathbf{A}) + H(\mathbf{B}) + H(\mathbf{C}) - H(\mathbf{AB}) - H(\mathbf{AC}) - H(\mathbf{BC}) + H(\mathbf{ABC}), \quad (33)$$

The above expression can be straightforwardly generalized for a larger number of probability schemes. For example, in the case of four dependent probability distributions of Scheme 5 one finds [6]:

$$I(\mathbf{A}:\mathbf{B}:\mathbf{C}:\mathbf{D}) = I(\mathbf{A}:\mathbf{B}:\mathbf{C}) - I(\mathbf{A}:\mathbf{B}:\mathbf{C}|\mathbf{D})$$

$$= \sum_{a\in\boldsymbol{a}}\sum_{b\in\boldsymbol{b}}\sum_{c\in\boldsymbol{c}}\sum_{d\in\boldsymbol{d}} P(a\wedge b\wedge c\wedge d)\log\frac{P(d|a\wedge b\wedge c)P(d|a)P(d|b)P(d|c)}{P(d)P(d|a\wedge b)P(d|b\wedge c)P(d|a\wedge c)}$$

$$= H(\mathbf{A}) + H(\mathbf{B}) + H(\mathbf{C}) + H(\mathbf{D})$$

$$- H(\mathbf{AB}) - H(\mathbf{AC}) - H(\mathbf{BC}) - H(\mathbf{AD}) - H(\mathbf{BD}) - H(\mathbf{CD})$$

$$+ H(\mathbf{ABC}) + H(\mathbf{ABD}) + H(\mathbf{ACD}) + H(\mathbf{BCD})$$

$$- H(\mathbf{ABCD}). \quad (34)$$

The conditional entropy of Eq. (28),

$$H(\mathbf{A}|\mathbf{BC}) \equiv H(\mathbf{A}|\mathbf{D}) = H(\mathbf{AD}) - H(\mathbf{D})$$

$$= S(\boldsymbol{P}(\boldsymbol{a})|\mathbf{P}(\boldsymbol{b}\wedge\boldsymbol{c})) \equiv S(\boldsymbol{P}(\boldsymbol{a})|\boldsymbol{P}(\boldsymbol{d})), \quad (35)$$

where the “product” probability scheme $\mathbf{D} = \mathbf{BC} = [\boldsymbol{d}, \boldsymbol{P}(\boldsymbol{d})] = [\boldsymbol{b}\wedge\boldsymbol{c}, \mathbf{P}(\boldsymbol{b}\wedge\boldsymbol{c})]$, measures the average noise in the $\boldsymbol{d}\rightarrow\boldsymbol{a}$ propagation of information in the underlying communication system. This entropy can be compared (see Scheme 4) with the two-scheme relative entropy $H(\mathbf{A}|\mathbf{B}) = H(\mathbf{A}) - I(\mathbf{A}:\mathbf{B})$, which measures the communication noise in the $\boldsymbol{b}\rightarrow\boldsymbol{a}$ probability scattering [see also Eq. (27)]:

$$H(\mathbf{A}|\mathbf{BC}) = -\sum_{a\in\boldsymbol{a}}\sum_{b\in\boldsymbol{b}}\sum_{c\in\boldsymbol{c}} P(a\wedge b\wedge c)\log\frac{P(a\wedge b\wedge c)}{P(b\wedge c)}$$

$$= -\sum_{a\in\boldsymbol{a}}\sum_{b\in\boldsymbol{b}}\sum_{c\in\boldsymbol{c}} P(a\wedge b\wedge c)\log\frac{P(a\wedge b\wedge c)p(a)}{p(a)P(b\wedge c)}$$

$$= H(\mathbf{A}) - I(\mathbf{A}:\mathbf{BC}) < H(\mathbf{A}|\mathbf{B}), \qquad (36)$$

since $I(\mathbf{A}:\mathbf{D}) > I(\mathbf{A}:\mathbf{B})$. The information propagation from the product events in the $\boldsymbol{d}\rightarrow\boldsymbol{a}$ channel thus results in less noise (IT-covalency) compared to that characterizing the $\boldsymbol{b}\rightarrow\boldsymbol{a}$ network. Therefore, the $\boldsymbol{d}\rightarrow\boldsymbol{a}$ channel appears to be more deterministic in character compared to $\boldsymbol{b}\rightarrow\boldsymbol{a}$ communication system.

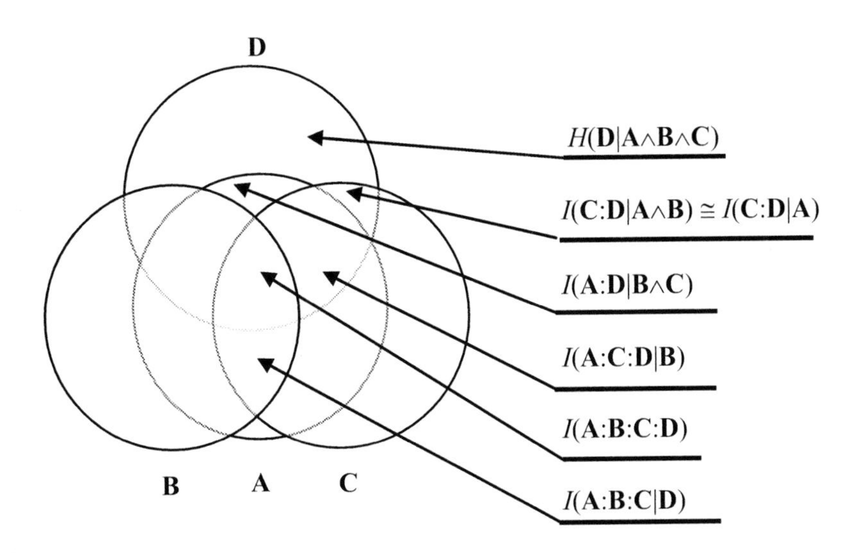

Scheme 5. General entropy/information diagrams of four probability schemes **A**, **B**, **C** and **D**. The entropies of the separate probability distributions are represented by circles, the circle overlap area denotes the relevant mutual-information quantity, while the circle remainder, after removal of their overlap(s) with other circle(s), signifies the corresponding conditional-entropy descriptor

Accordingly, the opposite trend must be detected in the complementary mutual-information quantity of Eq. (27):

$$I(\mathbf{A}:\mathbf{BC}) = H(\mathbf{A}) - H(\mathbf{A}|\mathbf{D}) > I(\mathbf{A}:\mathbf{B}) = H(\mathbf{A}) - H(\mathbf{A}|\mathbf{B}). \tag{37}$$

Therefore, the product-event scattering preserves in **D** a larger portion of the original information content of **A**, compared to that conserved in the scheme **B** alone, due to the extra information in **C** about **A** given **B**, as measured by $I(\mathbf{A}:\mathbf{C}|\mathbf{B})$. In other words, the $\boldsymbol{d}\to\boldsymbol{a}$ probability propagation is more deterministic in character, thus resulting in a larger IT-ionic component compared to that in the $\boldsymbol{b}\to\boldsymbol{a}$ scattering.

To summarize, the $(\boldsymbol{d} = \boldsymbol{b}\wedge\boldsymbol{c})\to\boldsymbol{a}$ communications appear less noisy (more deterministic) campared to the $\boldsymbol{b}\to\boldsymbol{a}$ probability propagation. Hence, the *a posteriori* indeterminacy of **A** relative to **B**, $H(\mathbf{A}|\mathbf{B})$, exceeds the *double a posteriori* entropy $H(\mathbf{A}|\mathbf{BC})$, which measures the indeterminacy of **A** relative to **B** and **C**, while the information about **A** provided by **B** alone, $I(\mathbf{A}:\mathbf{B})$, is lower than that provided jointly by **B** and **C**, $I(\mathbf{A}:\mathbf{BC})$, by the extra amount of information $I(\mathbf{A}:\mathbf{C}|\mathbf{B})$ about **A** provided by **C**, when **B** is known beforehand.

Let us finally examine the remaining *four*-scheme entropy/information descriptors depicted in Scheme 5. In this qualitative diagram we have delineated areas representing various indices reflecting the entropy/information couplings between the four dependent probability schemes, $\{\mathbf{X} = (\boldsymbol{x}, \boldsymbol{P}(\boldsymbol{x})\} = (\mathbf{A}, \mathbf{B}, \mathbf{C}, \mathbf{D})$, which combine the relevant sets of events, $\{\boldsymbol{x}\} = (\boldsymbol{a}, \boldsymbol{b}, \boldsymbol{c}, \boldsymbol{d})$, and their associated probabilities, $\{\boldsymbol{P}(\boldsymbol{x})\} = [\boldsymbol{P}(\boldsymbol{a}), \boldsymbol{P}(\boldsymbol{b}), \boldsymbol{P}(\boldsymbol{c}), \boldsymbol{P}(\boldsymbol{d})]$. Again, the Shannon entropies of the latter, $\{H(\mathbf{X}) = H[\boldsymbol{P}(\boldsymbol{x})]\}$, are depicted by circles. For example, these partial events may refer to finding in the bond system of the molecule an electron on the basis functions $\{\boldsymbol{x} = \chi_{\mathrm{X}}\}$ contributed by molecular fragments $\{\mathrm{X}\}$. In these diagrams the overlap areas between circles reflect the average mutual-information quantities, measuring the associated IT-ionicities, while the corresponding circle-remainders represent the complementary average conditional entropies, reflecting the relevant IT-covalency components. It should again be emphasized at this point that some of these descriptors may assume negative values.

Consider, e.g., the conditional entropy of **D**, given the simultaneous occurrence of events in **A**, **B**, and **C** [22],

$$H(\mathbf{D}\,|\,\mathbf{A}\wedge\mathbf{B}\wedge\mathbf{C}) = S[\boldsymbol{P}(\boldsymbol{d})|\boldsymbol{P}(\boldsymbol{a})\wedge\boldsymbol{P}(\boldsymbol{b})\wedge\boldsymbol{P}(\boldsymbol{c})]$$

$$= -\sum_{a\in\boldsymbol{a}}\sum_{b\in\boldsymbol{b}}\sum_{c\in\boldsymbol{c}}\sum_{d\in\boldsymbol{d}} P(a\wedge b\wedge c\wedge d)\log P(d|a\wedge b\wedge c). \tag{38}$$

It represents the residual uncertainty in **D**, when the events of the three remaining schemes are known to have occurred already. In the molecular fragment scenario it measures the overall IT-covalency of all bonds in fragment **D**, due to the basis functions of the remaining subsystems A, B, and C. The complementary IT-ionicity index reads [6,22]:

$$I(\mathbf{A}\wedge\mathbf{B}\wedge\mathbf{C}:\mathbf{D}) = I[\boldsymbol{P}(\boldsymbol{a})\wedge\boldsymbol{P}(\boldsymbol{b})\wedge\boldsymbol{P}(\boldsymbol{c}):\boldsymbol{P}(\boldsymbol{d})] = H(\mathbf{D}) - H(\mathbf{D}\,|\,\mathbf{A}\wedge\mathbf{B}\wedge\mathbf{C})$$

$$= \sum_{a\in\boldsymbol{a}}\sum_{b\in\boldsymbol{b}}\sum_{c\in\boldsymbol{c}}\sum_{d\in\boldsymbol{d}} P(a\wedge b\wedge c\wedge d)\log\frac{P(a\wedge b\wedge c\wedge d)}{P(d)P(a\wedge b\wedge c)}$$

$$= \sum_{a\in\boldsymbol{a}}\sum_{b\in\boldsymbol{b}}\sum_{c\in\boldsymbol{c}}\sum_{d\in\boldsymbol{d}} P(a\wedge b\wedge c\wedge d)\log\frac{P(d|a\wedge b\wedge c)}{P(d)}, \qquad (39)$$

thus giving rise to the overall (conditional) bond index in **D**:

$$N(\mathbf{A}\wedge\mathbf{B}\wedge\mathbf{C};\mathbf{D}) = H(\mathbf{D}\,|\,\mathbf{A}\wedge\mathbf{B}\wedge\mathbf{C}) + I(\mathbf{A}\wedge\mathbf{B}\wedge\mathbf{C}:\mathbf{D}) = H(\mathbf{D}). \qquad (40)$$

Using the property of the information *additivity* [6] one can alternatively express the amount of information of Eq. (39) in terms of more elementary mutual information quantities generating the overlap area between the envelope of three overlapping circles (**A**, **B**, **C**) and **D** (see Scheme 5):

$$I(\mathbf{A}\wedge\mathbf{B}\wedge\mathbf{C}:\mathbf{D}) = I(\mathbf{A}:\mathbf{D}) + I(\mathbf{B}:\mathbf{D}\,|\,\mathbf{A}\wedge\mathbf{C}) + I(\mathbf{C}:\mathbf{D}\,|\,\mathbf{A}\wedge\mathbf{B})$$

$$= I(\mathbf{A}:\mathbf{D}\,|\,\mathbf{B}\wedge\mathbf{C}) + I(\mathbf{B}:\mathbf{D}\,|\,\mathbf{A}\wedge\mathbf{C}) + I(\mathbf{C}:\mathbf{D}\,|\,\mathbf{A}\wedge\mathbf{B})$$

$$+ I(\mathbf{A}:\mathbf{C}:\mathbf{D}\,|\,\mathbf{B}) + I(\mathbf{A}:\mathbf{B}:\mathbf{D}\,|\,\mathbf{C}) + I(\mathbf{A}:\mathbf{B}:\mathbf{C}:\mathbf{D}). \qquad (41)$$

In conclusion of this section we briefly summarize the key expressions for the complementary communication-*noise* (IT-covalency) and information-flow (IT-ionicity) descriptors involving several probability distributions. We first recall that for *two* dependent probability schemes these basic quantities read:

$$H(\mathbf{B}\,|\,\mathbf{A}) = -\sum_{a\in\boldsymbol{a}}\sum_{b\in\boldsymbol{b}} P(a\wedge b)\log P(b|a)$$

$$= H(\mathbf{A}\wedge\mathbf{B}) - H(\mathbf{A}), \qquad (42)$$

$$I(\mathbf{A}:\mathbf{B}) = \sum_{a\in\boldsymbol{a}}\sum_{b\in\boldsymbol{b}} P(a\wedge b)\log\frac{P(b|a)}{P(b)} = H(\mathbf{B}) - S(\mathbf{B}\,|\,\mathbf{A})$$

$$= \sum_{a\in\boldsymbol{a}}\sum_{b\in\boldsymbol{b}} P(a\wedge b)\log\frac{P(a|b)}{P(a)} = H(\mathbf{A}) - S(\mathbf{A}\,|\,\mathbf{B})$$

$$= H(\mathbf{A}) + H(\mathbf{B}) - H(\mathbf{A}\wedge\mathbf{B}). \tag{43}$$

The associated *three*-scheme indices can be similarly expressed by the relevant information/entropy descriptors of the probabilities involved [6,22]:

$$I(\mathbf{A}:\mathbf{B}:\mathbf{C}) = I(\mathbf{A}:\mathbf{B}) - I(\mathbf{A}:\mathbf{B}\,|\,\mathbf{C})$$

$$= H(\mathbf{A}) + H(\mathbf{B}) + H(\mathbf{C})$$

$$- H(\mathbf{A}\wedge\mathbf{B}) - H(\mathbf{A}\wedge\mathbf{C}) - H(\mathbf{B}\wedge\mathbf{C})$$

$$+ H(\mathbf{A}\wedge\mathbf{B}\wedge\mathbf{C})$$

$$= \sum_{a\in\boldsymbol{a}}\sum_{b\in\boldsymbol{b}}\sum_{c\in\boldsymbol{c}} P(a\wedge b\wedge c)\log\frac{P(c|a)P(c|b)}{P(c)P(c|a\wedge b)}, \tag{44}$$

$$I(\mathbf{A}:\mathbf{B}\,|\,\mathbf{C}) = H(\mathbf{A}\,|\,\mathbf{C}) - H(\mathbf{A}\,|\,\mathbf{B}\wedge\mathbf{C})$$

$$= \sum_{a\in\boldsymbol{a}}\sum_{b\in\boldsymbol{b}}\sum_{c\in\boldsymbol{c}} P(a\wedge b\wedge c)\log\frac{P(a|b\wedge c)}{P(a|c)}, \tag{45}$$

For the *four* dependent probability distributions one similarly finds:

$$I(\mathbf{A}:\mathbf{B}:\mathbf{C}:\mathbf{D}) = I(\mathbf{A}:\mathbf{B}:\mathbf{C}) - I(\mathbf{A}:\mathbf{B}:\mathbf{C}\,|\,\mathbf{D})$$

$$= \sum_{a\in\boldsymbol{a}}\sum_{b\in\boldsymbol{b}}\sum_{c\in\boldsymbol{c}}\sum_{d\in\boldsymbol{d}} P(a\wedge b\wedge c\wedge d)\log\frac{P(d|a\wedge b\wedge c)P(d|a)P(d|b)P(d|c)}{P(d)P(d|a\wedge b)P(d|b\wedge c)P(d|a\wedge c)}, \tag{46}$$

$$I(\mathbf{A}:\mathbf{B}:\mathbf{C}\,|\,\mathbf{D}) = I(\mathbf{A}:\mathbf{B}\,|\,\mathbf{D}) - I(\mathbf{A}:\mathbf{B}\,|\,\mathbf{C}\wedge\mathbf{D}), \tag{47}$$

where the conditional mutual-information measure

$$I(\mathbf{A}{:}\mathbf{B} \mid \mathbf{C}\wedge\mathbf{D}) = H(\mathbf{A} \mid \mathbf{C}\wedge\mathbf{D}) - H(\mathbf{A} \mid \mathbf{B}\wedge\mathbf{C}\wedge\mathbf{D})$$

$$= \sum_{a\in\boldsymbol{a}}\sum_{b\in\boldsymbol{b}}\sum_{c\in\boldsymbol{c}}\sum_{d\in\boldsymbol{d}} P(a\wedge b\wedge c\wedge d)\log\frac{P(a|b\wedge c\wedge d)}{P(a|c\wedge d)}. \tag{48}$$

1.6. Variational Principles

Let us recall some properties of the Shannon entropy, which should be expected from a reasonable measure of uncertainty/information. For the simplest case of a finite, discrete probability distribution $\boldsymbol{p} = \{p_i\}$, $S(\boldsymbol{p}) = -\sum_i p_i \log p_i = 0$ if and only if one of the probabilities in $\boldsymbol{p}$ is one and all the others are zero. Indeed, this is the case, where the outcome of the experiment carries no uncertainty, since the result can be predicted beforehand with complete certainty. In all other cases $S(\boldsymbol{p}) > 0$. For the fixed number of n outcomes the probability scheme exhibiting the most uncertainty, i.e., the largest value of the entropy, is the one with equally likely outcomes: $p_i = 1/n$. This "maximum smoothness" property of the optimum probability distribution is summarized by the constrained variational principle of Shannon and Jaynes, called the *Maximum Entropy* (ME) rule:

> given data $\boldsymbol{F}^0 = \{F_i^0\}$, the probability distribution $\boldsymbol{p}$ [or $p(\boldsymbol{r})$], which describes these constraints most objectively, must maximize the entropy $S(\boldsymbol{p})$ (or $S[p]$) with respect to all $\boldsymbol{p}$'s [$p(\boldsymbol{r})$'s] satisfying $\boldsymbol{F}^0$.

This implies the associated Euler-Lagrange variational principle:

$$\delta\{S(\boldsymbol{p}) - \sum_i \lambda_i[F_i(\boldsymbol{p}) - F_i^0]\} = \delta\{S(\boldsymbol{p}) - \sum_i \lambda_i F_i(\boldsymbol{p})\} = 0$$

or

$$\delta\{S[p] - \sum_i \lambda_i F_i[p]\} = 0, \tag{49}$$

where $\{\lambda_i\}$ denote the Lagrange multipliers enforcing the relevant auxiliary conditions $\{F_i(\boldsymbol{p}) = F_i^0$ or $F_i[p] = F_i^0\}$, all to be eventually determined from the constraint values. Thus, the entropy maximization results in the most "evenly spread" of all distributions consistent with the imposed constraints. This principle represents a device allowing one to assimilate in the optimum probability distribution the physical information contained in the applied constraints in the most unbiased manner possible.

The ME principle represents a powerful method for determining the equilibrium probability distribution of a physical system or process, given some information about it. It has been successfully used to reconstruct the equilibrium thermodynamics, asserting at the same time the principle applicability to far more general problems. The ME rule involves the maximization of Shannon's entropy subject to the imposed constraints and provides a unifying principle in statistical physics, allowing the construction of thermodynamic laws based upon statistical inference and the unbiased assimilation of available data. In other words, the entropy becomes the starting point in the construction of the statistical mechanics, instead of being identified in the end as a byproduct of the energy-centered arguments. This Shannon-Jaynes principle is the uniquely correct method for inductive inference, when a new information is given in the form of the statistical expectation values.

Kullback has proposed an important generalization of the ME rule to problems involving the reference probability distribution(s), which can be called the *Maximum-Resemblance* principle or the *Minimum Entropy Deficiency* (MED) rule, as another unbiased procedure for simultaneously assimilating in the optimum probability the constraints of available experimental data and its highest similarity to the reference distribution afforded by the constraints. Suppose that an experiment has been performed, which yields the expectation values of several functions of the discrete probability distribution $\boldsymbol{p}$, $\{F_i(\boldsymbol{p}) = F_i^0\}$, or functionals of the probability density $p(\boldsymbol{r})$ in the continuous case, $\{F_i[p] = F_i^0\}$. The Kullback principle asserts that the task of assimilating this information in the optimum probability distribution, which is to resemble the reference distribution $\boldsymbol{p}^0$ [or $p^0(\boldsymbol{r})$] as much as possible, can be accomplished by minimizing the entropy deficiency subject to these constraints:

$$\delta\{\Delta S(\boldsymbol{p}\,|\,\boldsymbol{p}^0) - \Sigma_i\, \lambda_i\, F_i(\boldsymbol{p})\} \equiv \delta\{\Delta S(\boldsymbol{p}\,|\,\boldsymbol{p}^0) - J(\boldsymbol{p})\} = 0$$

or

$$\delta\{\Delta S[p\,|\,p^0] - \Sigma_i\, \lambda_i\, F_i[p]\} \equiv \delta\{\Delta S[p\,|\,p^0] - J[p]\} = 0. \qquad (50)$$

Alternatively, the Fisher measure of information, which reflects the overall "*order*" ("sharpness") of the probability distribution, can be used in the variational procedure of assimilating the available data in the continuous probability distribution:

$$\delta\{I[p] - \Sigma_i \lambda_i F_i[p]\} \equiv \delta\{I[p] - J[p]\} \equiv \delta K[p] = 0. \tag{51}$$

This variational principle for the continuous probability density $p(\boldsymbol{r})$ contains two additive components of the *physical information* functional $K[p]$: the Fisher *intrinsic* information $I[p]$ and the *bound* (constrained) information $J[p]$, which usually represents the information associated with specific physical parameters, thus characterizing the effect of the measurement process [2]. Notice, that at the optimum solution $\delta I[p] = \delta J[p]$.

This principle of the *Extreme* (minimum) *Physical Information* (EPI) is closely related to the theory of measurement [2]. The maximum Shannon entropy and the minimum Fisher information principles, for a given set of constraints, may have coincident solutions. The Maxwell-Boltzmann velocity dispersion law of the equilibrium statistical thermodynamics results from both these information principles, with the Fisher information rule generating additional, non-equilibrium solutions as subsidiary minima, with the absolute minimum being attained by the Maxwell-Boltzmann solution. The coincidence is observed at the equilibrium level of statistical mechanics, with the Shannon ME rule being unable to cover the non-equilibrium phenomena. We shall briefly argue in the next section that the differential Schrödinger equations (SE) of quantum mechanics (stationary and time-dependent) also result from the EPI principle using the intrinsic Fisher information measure, as do the electronic and nuclear SE of the adiabatic (BO) separation of electronic (fast) and nuclear (slow) motions in molecular systems.

Chapter 2

SCHRÖDINGER EQUATIONS FROM FISHER INFORMATION

2.1. FISHER INFORMATION FOR COMPLEX PROBABILITY AMPLITUDES

For simplicity, let us consider a single spin-less particle of mass μ, characterized by the Hamiltonian (energy) operator

$$\hat{\mathrm{H}}(\boldsymbol{r}) = -(\hbar^2/2\mu)\nabla^2 + v(\boldsymbol{r}) = \hat{\mathrm{T}}(\boldsymbol{r}) + \hat{\mathrm{V}}(\boldsymbol{r}), \tag{52}$$

with $\hat{\mathrm{V}}(\boldsymbol{r}) \equiv v(\boldsymbol{r})$ standing for the (multiplicative) potential energy operator, e.g., of the nuclear-electron attraction energy, $v = v_{ne}$, and the Laplacian operator $\Delta = \nabla^2$. In the Schrödinger's mechanics the system quantum state is specified by the complex wave-function in the position representation,

$$\psi(\boldsymbol{r}, t) = R(\boldsymbol{r}, t) \exp[i\Phi(\boldsymbol{r}, t)], \tag{53}$$

where the real functions $R(\boldsymbol{r}, t)$ and $\Phi(\boldsymbol{r}, t)$ describe the probability-amplitude and the phase of ψ, respectively. Its physical interpretation is established through the particle probability density defined by its amplitude $R(\boldsymbol{r}, t)$,

$$p(\boldsymbol{r}, t) = |\psi(\boldsymbol{r}, t)|^2 = \psi^*(\boldsymbol{r}, t)\, \psi(\boldsymbol{r}, t) = R^2(\boldsymbol{r}, t), \qquad \int p(\boldsymbol{r}, t)\, d\boldsymbol{r} = 1, \tag{54}$$

and the *probability-current* density:

$$\boldsymbol{j}(\boldsymbol{r}, t) = \frac{\hbar}{2\mu i}[\psi^*(r,t)\nabla\psi(r,t) - \psi(r,t)\nabla\psi^*(r,t)] \equiv p(\boldsymbol{r}, t)\, \boldsymbol{V}(\boldsymbol{r}, t)$$

$$= \frac{\hbar}{\mu}\operatorname{Im}(\psi^*\nabla\psi) \equiv \boldsymbol{j}[\psi] = \frac{\hbar}{\mu} p\nabla\Phi = p\nabla\left[\frac{\hbar\Phi}{\mu}\right]. \tag{55}$$

The preceding equation also expresses the local speed of the probability "fluid", $\boldsymbol{V}(\boldsymbol{r}, t) = \boldsymbol{j}(\boldsymbol{r}, t)/p(\boldsymbol{r}, t)$, in terms of the gradient of the phase-part of the wave-function. The probability-current distribution is seen to explore the gradient of the phase function weighted by the local probability density. The latter is determined solely by the density-amplitude, while the speed of the probability fluid is proportional to the gradient of the phase part of the wave-function.

For such complex probability amplitudes of quantum mechanics the classical Fisher measure of information of Eq. (5), $I[p] = \mathrm{I}[R]$, is generalized into the functional of Eq. (6) with its integrand defining the associated information density $f(\boldsymbol{r}, t)$:

$$I[\psi] = 4\int|\nabla\psi(\boldsymbol{r}, t)|^2 d\boldsymbol{r} = 4\int\nabla\psi^*(\boldsymbol{r}, t)\cdot\nabla\psi(\boldsymbol{r}, t)\, d\boldsymbol{r} \equiv \int f(\boldsymbol{r}, t)\, d\boldsymbol{r}. \tag{56}$$

Its analog for the multi-component systems, described by the vector of component wave-functions $\boldsymbol{\psi} = \{\psi_n\}$, then reads:

$$I[\boldsymbol{\psi}] = 4\sum_n\int|\nabla\psi_n(\boldsymbol{r}, t)|^2 d\boldsymbol{r} = \sum_n I[\psi_n] \equiv \sum_n\int f_n(\boldsymbol{r}, t)\, d\boldsymbol{r}. \tag{57}$$

In what follows we shall examine some properties of this generalized information measure. We first observe that by a straightforward integration by parts the Schrödinger functional for the expectation value of the kinetic energy operator can be interpreted as being proportional to the average generalized Fisher information contained in the wave-function ψ:

$$\langle T\rangle = \int\psi^*\hat{\mathrm{T}}\psi\, d\boldsymbol{r} \equiv \langle\psi|\hat{\mathrm{T}}|\psi\rangle$$

$$= \frac{\hbar^2}{2\mu}\int\nabla\psi^*\cdot\nabla\psi\, d\boldsymbol{r} = \frac{\hbar^2}{8\mu}I[\psi]\cdot \tag{58}$$

In terms of the probability-amplitude and phase parts of the system wave-function this Fisher information functional reads:

$$I[\psi] = 4\int\{(\nabla R)^2 + P\,(\nabla \Phi)^2\}\; d\boldsymbol{r} \equiv I[R] + I[R,\,\Phi]$$

$$= \int \frac{(\nabla p)^2}{p}\, d\boldsymbol{r} + \frac{4\mu^2}{\hbar^2}\int \frac{\boldsymbol{j}^2}{p}\, d\boldsymbol{r} \equiv I[p] + I[\boldsymbol{j}] \equiv I[p, \boldsymbol{j}]. \tag{59}$$

Therefore, this generalized measure of information becomes identical with the classical Fisher functional $I[p]$ for the *stationary* quantum states which are characterized by the time-independent probability amplitude $R = \varphi(\boldsymbol{r})$ and the position-independent phase $\Phi = -\omega t$:

$$\psi(\boldsymbol{r}, t) = \varphi(\boldsymbol{r}) \exp[-iEt/\hbar] \equiv \varphi(\boldsymbol{r}) \exp[-i\omega t]. \tag{60}$$

These eigenfunctions of the Hamiltonian operator,

$$\hat{\mathrm{H}}\psi = E\psi \qquad \text{or} \qquad \hat{\mathrm{H}}\varphi = E\varphi, \tag{61}$$

correspond to the sharply-specified system energy E and the frequency ω. Indeed, such states imply the vanishing current, $\boldsymbol{j} = 0$, and hence also $I[\boldsymbol{j}] = 0$.

The generalized functional $I[\psi]$ is seen to symmetrically probe the gradient content of both parts of the complex wave-function, with the gradient of the probability-amplitude determining the classical Fisher measure of information contained in the probability distribution p, with the latter also providing the weight in the local contribution from the phase gradient, due to the probability-current density,

$$I[p] \equiv I^F[R] = \int\left(\frac{\nabla p}{\sqrt{p}}\right)^2 d\boldsymbol{r} \equiv \int(\overline{\nabla} p)^2\, d\boldsymbol{r}, \tag{62}$$

$$I[\boldsymbol{j}] = \int(2R\nabla\Phi)^2\, d\boldsymbol{r} \equiv I[R,\,\Phi]$$

$$= \int\left(\frac{2\mu\,\boldsymbol{j}}{\hbar\sqrt{p}}\right)^2 d\boldsymbol{r} \equiv \int \bar{\boldsymbol{j}}^2\, d\boldsymbol{r}\,. \tag{63}$$

It follows from these expressions that the classical Fisher information measures the "length" of the "reduced" gradient of the probability density, $\overline{\nabla} p(\boldsymbol{r})$, while the other contribution represents the length of the reduced vector of the probability-current density $\bar{j}(r)$. These two information contributions can be alternatively expressed in terms of the real and imaginary parts of the gradient of the wave-function logarithm, $\nabla \ln \psi = (\nabla \psi)/\psi$,

$$I[R] = 4 \int p[\mathrm{Re}(\nabla \ln \psi)]^2 \, d\boldsymbol{r} \quad \text{and} \quad I[R, \Phi] = 4 \int p[\mathrm{Im}(\nabla \ln \psi)]^2 \, d\boldsymbol{r}. \tag{64}$$

Thus, these complementary components of the generalized Fisher information thus have a common interpretation in quantum mechanics as the p-weighted averages of the gradient content of the real and imaginary parts of the logarithmic gradient of the system wave-function. As such they represent a natural (complex) generalization of the classical (*real*) information concept of Eq. (5).

Of interest also is the information density per electron:

$$\tilde{f} \equiv \frac{f}{p} = \left(\frac{\nabla p}{p}\right)^2 + \left(\frac{2\mu \boldsymbol{j}}{\hbar p}\right)^2 \equiv (\tilde{\nabla} p)^2 + (\tilde{\boldsymbol{j}})^2 \geq 0. \tag{65}$$

It is seen to be generated by the squares of the local values of the related quantities per electron: the probability gradient $(\tilde{\nabla} p)^2$ and the current density $(\tilde{\boldsymbol{j}})^2$. This expression emphasizes the basic equivalence of the roles played by the probability density and its current in shaping the resultant value of the generalized Fisher-information density in quantum mechanics.

2.2. Continuity Equations

For the IT interpretation of electron redistributions in molecules the concept of the information *flow* is paramount. The Hamiltonian operator $\hat{\mathrm{H}}$ determines the time-evolution of the system wave-function through the Schrödinger equation

$$i\hbar \frac{\partial \psi}{\partial t} = \hat{\mathrm{H}} \psi, \tag{66}$$

which implies the conservation of the wave-function normalization in time:

$$\frac{d}{dt} \int \psi^*(\boldsymbol{r},t)\, \psi(\boldsymbol{r},t)\, d\boldsymbol{r} = \frac{d}{dt} \int p(\boldsymbol{r},t)\, d\boldsymbol{r} = 0. \tag{67}$$

The probability density and its current together determine the local balance of the probability distribution in quantum mechanics which is summarized by the familiar *continuity equation*:

$$\frac{\partial p}{\partial t} = -\nabla \cdot \boldsymbol{j} \qquad \text{or} \qquad \frac{dp}{dt} \equiv \dot{p} = \frac{\partial p}{\partial t} + \nabla \cdot \boldsymbol{j} = 0 . \tag{68}$$

The first form of the preceding equation expresses the fact that for the norm conserving evolution of the system wave-function the local rate of change of the probability density is determined by the probability density leaving that location, so that the local net production (surce) $\dot{p}$ of the probability density identically vanishes, as expressed by the second form of this equation. Indeed the particles are neither created nor destroyed in (closed) molecular systems.

Moreover, by Green's theorem, the volume integral representing the overall outflow of the particle probability from the volume V enclosed by the closed surface S can be expressed as the surface integral measuring the global flux through S:

$$\int_V \nabla \cdot \boldsymbol{j}\, d\boldsymbol{r} = \int_S \boldsymbol{j} \cdot d\boldsymbol{S} , \tag{69}$$

with $d\boldsymbol{S} = dS\boldsymbol{n}$ standing for the normal vector of the surface element dS along the unit vector $\boldsymbol{n}$. By using Eq. (69) the generalized Fisher information of Eq. (59) can be expressed as the following difference between the relevant surface and volume integrals,

$$\begin{aligned} I[\psi] &= 4\int \nabla \psi^* \cdot \nabla \psi\, d\boldsymbol{r} \\ &= 4\left[\int_S \psi^* (\nabla \psi) \cdot d\boldsymbol{S} - \int_V \psi^* \Delta \psi\, d\boldsymbol{r} \right], \end{aligned} \tag{70}$$

with $V \to \infty$, corresponding to a very large volume defined by the closed surface $S \to \infty$. Subtracting from this equation its complex conjugate and taking into account the Hermitian property of the Laplacian then gives the conservation of the overall probability in V:

$$\int_S \boldsymbol{j} \cdot d\boldsymbol{S} = 0 . \tag{71}$$

It should be stressed, however, that - with the exception of the stationary quantum states - the overall amount of information in the system is not generally conserved, when the particle probability density evolves in time. Therefore, the continuity equation expressing the local balance of the Fisher information in quantum mechanics has to include a non-vanishing "source" term $df/dt = \dot{f}$, since the overall information content, proportional to the system average kinetic energy, changes for different shapes of the probability density:

$$\frac{\partial f}{\partial t} = \dot{f} - \nabla \cdot \boldsymbol{J} \qquad \text{or} \qquad \frac{df}{dt} = \dot{f} = \frac{\partial f}{\partial t} + \nabla \cdot \boldsymbol{J} \neq 0,$$

$$\boldsymbol{J} = f\boldsymbol{v} = \left(\frac{f}{p}\right)\boldsymbol{j} = \tilde{f}\boldsymbol{j}. \tag{72}$$

Above, the information-current density $\boldsymbol{J}$ exhibits the same local velocity $\boldsymbol{v}$ as the probability "fluid" in Eq. (55), since the information current in the system is effected through the probability flow $\boldsymbol{j}$.

In the continuity Eqs. (68) and (72) the partial derivatives $\partial p/\partial t$ and $\partial f/\partial t$ measure the rate of change at the *fixed* point in space, inside the infinitesimal volume element at rest. Alternatively, the continuity equations have been interpreted in terms of the total derivatives (sources) $\dot{p}$ and $\dot{f}$, which represent the time rate of change of these densities in a volume element of the particle-probability fluid as it moves in space with the speed $\boldsymbol{v}$.

Therefore, in order to explicitly identify the density of information production we have to determine the total derivative df/dt (see also Appendix A). We first observe that the quantum-mechanical operator $\hat{\mathrm{I}}(\boldsymbol{r})$ (linear and Hermitian) of the generalized Fisher information, which defines its expectation value of Eq. (56),

$$I \equiv \langle I \rangle = \int \psi^*(\boldsymbol{r})\hat{\mathrm{I}}(\boldsymbol{r})\psi(\boldsymbol{r})\,d\boldsymbol{r}, \tag{73}$$

is proportional to the negative Laplacian $\Delta = \nabla^2$:

$$\hat{\mathrm{I}}(\boldsymbol{r}) = -4\Delta = 4(\hat{\mathbf{p}}(\boldsymbol{r})/\hbar)^2 \equiv 4\hat{\mathbf{k}}(\boldsymbol{r})^2, \tag{74}$$

where $\hat{\mathbf{p}}(\boldsymbol{r}) = -i\hbar\nabla \equiv \hbar\hat{\mathbf{k}}(\boldsymbol{r})$ stands for the quantum-mechanical operator of the particle momentum with $\hat{\mathbf{k}}(\boldsymbol{r})$ standing for the particle wave-vector operator. The average information $\langle I \rangle$ thus measures the average length of the particle momentum vector. Indeed, for the wave-functions and their gradients vanishing at infinity the identity of Eq. (73) follows directly via a straightforward integration by parts [see also Eq. (58)].

Furthermore, since $\hat{\mathrm{I}}(\boldsymbol{r})$ does not depend explicitly on time, the quantum-mechanical expression for the time rate of change of $\langle I \rangle$, which also directly follows from Eq. (66), is determined by the expectation value of the commutator $[\hat{\mathrm{H}}, \hat{\mathrm{I}}]$ (see Appendix A):

$$\frac{\partial \langle I \rangle}{\partial t} \equiv \int \frac{\partial f(\boldsymbol{r})}{\partial t} d\boldsymbol{r} = \frac{i}{\hbar}\left\langle [\hat{\mathrm{H}}, \hat{\mathrm{I}}] \right\rangle = \frac{i}{\hbar}\left\langle [\hat{\mathrm{V}}, \hat{\mathrm{I}}] \right\rangle. \tag{75}$$

This relation provides the physical interpretation of the time rate of change of the Fisher information density at the fixed location in space. It is seen to be determined by the commutator

$$[\hat{\mathrm{V}}, \hat{\mathrm{I}}] = 4[\Delta, v] = 8(\nabla v) \cdot \nabla + 4(\Delta v). \tag{76}$$

The total time derivative of the Fisher information density, $\dot{f}(\boldsymbol{r}) = df(\boldsymbol{r})/dt$, which determines the source term in the associated continuity equation, is discussed in detail in Appendix A, where it is shown to be proportional to the probability current density, locally weighted by the gradient of the external potential, which determines the negative mechanical force acting on the particle, and the gradient of the information density itself:

$$\dot{f}(\boldsymbol{r}) = \frac{df(\boldsymbol{r})}{dt} = \frac{\partial f(\boldsymbol{r})}{\partial t} + \frac{\nabla f}{p} \cdot \boldsymbol{j} = 4\left(\frac{2\mu}{\hbar^2}\nabla v + \frac{\nabla f}{p}\right) \cdot \boldsymbol{j}. \tag{77}$$

This local production of the Fisher information is thus given by the product of the particle flux $\boldsymbol{j}$ and the "information-force" (gradient) term, which resembles the associated expression for the local entropy source in the ordinary irreversible thermodynamics expressed as products of thermodynamic forces and the conjugated fluxes.

Finally, we would also like to recall that in a general case of the non-stationary quantum states, when the system average energy is not conserved, it is the stationary property of the quantum-mechanical action integral

$$\mathcal{A} = \int_{t_0}^{t_1} dt \langle \psi(t) | \hat{\mathrm{A}}(t) | \psi(t) \rangle, \quad . \tag{78}$$

with the expectation value $A(t) = \langle \psi(t) | \hat{\mathrm{A}}(t) | \psi(t) \rangle$ defined by the action operator $\hat{\mathrm{A}}(t) = i\hbar\partial/\partial t - \hat{\mathrm{H}}$, which generates the time-dependent Schrödinger equation. Indeed, Eq. (66) is seen to directly result from the *stationary-action principle* $\delta\mathcal{A} = 0$, which can be equivalently expressed by the vanishing functional derivative

$$\frac{\delta\mathcal{A}}{\delta\psi^*(\boldsymbol{r},t)} = 0 \cdot \tag{79}$$

It replaces in the time-dependent problem the energy variational principle of the stationary case.

2.3. Stationary Schrödinger Equation as Information Principle

Consider again the stationary wave-equation of quantum mechanics [Eq. (61)] for a general N-electron molecular system,

$$\hat{\mathrm{H}}(N)\Psi(N) = E\Psi(N). \tag{80}$$

It marks the eigenvalue equation of the electronic (fixed-nuclei) Hamiltonian in the familiar adiabatic (BO) approximation,

$$\hat{\mathrm{H}}(N) = -(\hbar^2/2m_e)\sum_{i=1}^{N}\Delta_i + V(N) = \hat{\mathrm{T}}(N) + \hat{\mathrm{V}}(N), \tag{81}$$

where the (multiplicative) potential-energy operator $\hat{\mathrm{V}}(N) = V_{ne}(N) + V_{ee}(N) = V(N)$ includes the nuclear-electron (ne) attraction as well as the electron-electron (ee) repulsion terms, and E stands for the system ground-state electronic energy.

This equation directly follows from the Schrödinger variational principle for the system minimum average energy $\langle E(N)\rangle$ subject to the normalization constraint of the electronic ground-state wave-function:

$$\delta\{\langle E(N)\rangle - E(N)\langle \Psi(N)|\Psi(N)\rangle\} = 0, \tag{82a}$$

where the exact electronic energy $E(N)$ plays the role of the Lagrange multiplier enforcing the constraint.

The average electronic energy, the expectation value of the Hamiltonian of Eq. (81), contains the kinetic and potential contributions,

$$\langle E(N)\rangle = \langle \Psi(N)|\hat{\mathrm{H}}(N)|\Psi(N)\rangle = \langle \Psi(N)|\hat{\mathrm{T}}(N)|\Psi(N)\rangle + \langle \Psi(N)|\hat{\mathrm{V}}(N)|\Psi(N)\rangle$$

$$= \langle T(N)\rangle + \langle V(N)\rangle, \tag{82b}$$

with the kinetic-energy $\langle T\rangle$ component being proportional to the N-electron Fisher information:

$$\langle I(N)\rangle = \int \sum_{i=1}^{N} \frac{|\nabla_i P(N)|^2}{P(N)} d\tau^N$$

$$= -4\sum_{i=1}^{N} \langle \Psi(N)|\Delta_i|\Psi(N)\rangle = 4\sum_{i=1}^{N} \langle \nabla_i \Psi^*(N) \cdot \nabla_i \Psi(N)\rangle = \frac{8m_e}{\hbar^2}\langle T(N)\rangle. \tag{83}$$

Here the joint-probability distribution of N electrons,

$$P(N) = \Psi^*(N)\,\Psi(N) = |\Psi(N)|^2, \qquad \int P(N) d\tau^N = 1, \tag{84}$$

determines the average value of the system potential energy,

$$\langle V(N)\rangle = \int P(N)\, V(N)\, d\tau^N. \tag{85}$$

Therefore, the wave-function principle of Eq. (82a) also expresses the stationary character of the constrained Fisher information in the associated EPI rule [2,7,24,37,38]:

$$\delta\left\{\langle I(N)\rangle + \frac{8m_e}{\hbar^2}\left[\int P(N)\,V(N)\,d\tau^N - E(N)\int P(N)\,d\tau^N\right]\right\}$$

$$= \delta\left(\langle I[P(N)]\rangle - \lambda_{\langle V\rangle}\langle V[P(N)]\rangle - \lambda_{Norm.}\int P(N)\,d\tau^N\right) = 0. \quad (86)$$

Here, the average Fisher information term $\langle I(N)\rangle$ represents the EPI *intrinsic*-information term and the remaining, constraint part, stands for the EPI *bound*-information contribution. The latter consists of two terms: the physical, potential-energy part, effected by the Lagrange multiplier $\lambda_{\langle V\rangle} = -8m_e/\hbar^2$, and the "geometric" condition of the system probability normalization enforced by $\lambda_{Norm.} = 8m_e E/\hbar^2$.

Therefore, the stationary Schrödinger equation determines the optimum wave-function (probability amplitude) which marks the extremum of the N-electron Fisher information subject to the probability normalization and the average potential energy constraints. Therefore, the time-independent Schrödinger equation can be alternatively regarded as having the information origins, by resulting from the EPI rule of Eq. (86). It is satisfied by the optimum probability distribution $P(N)$ of the electronic ground state. It should be also observed that the two Lagrange multipliers in this information principle can be also interpreted as corresponding derivatives of the optimum Fisher information with respect to the associated constraints [24].

2.4. Time-Dependent Schrödinger Equation from Fisher Information

Let us briefly examine sufficient physical constraints in the relevant Euler-Lagrange information principle giving rise to the time-dependent Schrödinger equation (66). For simplicity we again assume the simplest case of a single, spin-less particle of mass μ described by the wave function ψ [Eq. (53)]. As we have demonstrated in the preceding section the time-independent (stationary) Schrödinger equation results from the Euler-Lagrange EPI principle for the trial probability-amplitudes $R'(\boldsymbol{r}) = \varphi'(\boldsymbol{r})$ of Eq. (60), which determine the variational probability density $p'(\boldsymbol{r}) = |\varphi'(\boldsymbol{r})|^2$, including the constraint terms due to the probability/wave-function normalization [19],

$$\|\psi'\|^2 = \int \psi'^* \psi' d\boldsymbol{r} = \int p(\boldsymbol{r})\, d\boldsymbol{r} = 1, \tag{87}$$

and the fixed value of the average potential energy of the interaction between the electron density with the external potential $v(\boldsymbol{r})$ due to the fixed nuclei:

$$\langle V \rangle = \int \psi^*(\boldsymbol{r}) v(\boldsymbol{r}) \psi(\boldsymbol{r})\, d\boldsymbol{r} = \int p(\boldsymbol{r}) v(\boldsymbol{r})\, d\boldsymbol{r} = V^0, \tag{88}$$

These constraints have been found to be enforced by the global Lagrange multipliers $\alpha = 8Em_e/\hbar^2$ and $\beta = -8m_e/\hbar^2$, respectively,

$$\delta\{\, I[\psi'] - \alpha \int \psi'^*(\boldsymbol{r})\psi'(\boldsymbol{r})\, d\boldsymbol{r} - \beta \int \psi'^*(\boldsymbol{r}) v(\boldsymbol{r})\psi'(\boldsymbol{r})\, d\boldsymbol{r} \,\} = 0, \tag{89}$$

where E stands for the fixed ground-state energy of the one-electron system.

The *non*-stationary wave-equation (66) additionally requires [24] the time-dependent constraint. The stationary action principle of Eq. (79) for trial variations $\delta\psi^*(\boldsymbol{r},t) = \langle \delta\psi(t) | \boldsymbol{r} \rangle$,

$$\delta\mathfrak{a} = \int_{t_0}^{t_1} dt \langle \delta\psi(t) | \hat{\mathrm{A}}(t) | \psi(t) \rangle = i\hbar \int_{t_0}^{t_1} dt \langle \delta\psi(t) | \partial\psi(t)/\partial t \rangle - \int_{t_0}^{t_1} dt \langle \delta\psi(t) | \hat{\mathrm{H}} | \psi(t) \rangle,$$

suggests that the associated non-stationary EPI rule has to include the time derivative of the squared norm of Eq. (87),

$$\frac{\partial \|\psi'\|^2}{\partial t} = \int \left(\frac{\partial \psi'^*}{\partial t} \psi' + \psi'^* \frac{\partial \psi'}{\partial t} \right) d\boldsymbol{r} = \int \frac{\partial p(\boldsymbol{r})}{\partial t} d\boldsymbol{r},$$

besides the variation of the system average electronic energy. Threfore, the conservation of the probability normalization in time [Eq. (67)],

$$\frac{\partial \langle \psi | \psi \rangle}{\partial t} = \frac{\partial}{\partial t} \int p(\boldsymbol{r}) d\boldsymbol{r} = 0, \tag{90}$$

provides the relevant time-dependent constraint in the EPI principle generating the Schrödinger dynamics of quantum states.

For the normalized wave-functions, when $\alpha = 0$ (redundant constraint), the corresponding Fisher-information EPI rule then reads:

$$\delta\left\{I[\psi'] - \beta\int\psi'^*(\boldsymbol{r})v(\boldsymbol{r})\psi'(\boldsymbol{r})d\boldsymbol{r} - \gamma\frac{\partial}{\partial t}\int\psi'^*(\boldsymbol{r})\psi'(\boldsymbol{r})d\boldsymbol{r}\right\} \equiv \delta\mathcal{J}[\psi'^*,\psi'] = 0, \qquad (91)$$

where β has already been identified before and $\gamma = 8\mu i/\hbar$. This identification follows from the functional differentiation of the above auxiliary functional $\mathcal{J}[\psi'^*,\psi']$ with respect to ψ'^*. These Lagrange multipliers can be also interpreted as derivatives of the optimum Fisher information with respect to the associated constraints [24].

2.5. Information Principle Generating Kohn-Sham Equations

We shall now briefly demonstrate that the familiar *Kohn-Sham* (KS) equations [40] of the computational *Density Functional Theory* (DFT) [41-44] also result from the associated EPI principle [7]. To simplify the notation the *atomic units* (a.u.) will be used. The relevant intrinsic-data information-functional for this orbital approximation is now in the *multi*-component Fisher form $I[\boldsymbol{\psi}]$ [Eq. (57)], where $\boldsymbol{\psi} = \{\psi_n\}$ groups the N-lowest *singly*-occupied KS *spin*-MO. Indeed, in KS theory each orbital probability density $\rho_n(\boldsymbol{r}) = |\psi_n(\boldsymbol{r})|^2$ constitutes a distinct component of the overall *one*–electron probability distribution $p(\boldsymbol{r}) = \rho(\boldsymbol{r})/N$, where the ground-state electron density $\rho(\boldsymbol{r})$ of the fictitious non-interacting system of N particles is by hypothesis equal to that of the real system of the fully interacting electrons:

$$\rho(\boldsymbol{r}) = \Sigma_{\sigma=\uparrow,\downarrow}\Sigma_n\,|\psi_n(\boldsymbol{r},\sigma)|^2 \equiv \Sigma_n\,\rho_n(\boldsymbol{r}) \equiv \Sigma_{\sigma=\uparrow,\downarrow}\rho_\sigma(\boldsymbol{r}),$$

$$\int\rho(\boldsymbol{r})\,d\boldsymbol{r} = N, \qquad \int\rho_\sigma(\boldsymbol{r})\,d\boldsymbol{r} = N_\sigma; \qquad (92)$$

here $\{\rho_\sigma(\boldsymbol{r})\}$ are the two spin densities and N_σ denotes the number of electrons exhibiting the spin-up ($\uparrow$, $\sigma = ½$) or the spin-down ($\downarrow, \sigma = -½$) states of an electron. The physical information functional $K[\boldsymbol{\psi}] = I[\boldsymbol{\psi}] - J[\boldsymbol{\psi}]$ of the associated

EPI principle [Eq. (51)], where $J[\psi]$ stands for the *multi*-component constrained (bound) information part, then uniquely defines the variational KS problem for the molecular system in question.

Let us now consider in some detail the key *one*-body problem of KS theory of N *non*-interacting electrons moving in an effective (local) external potential $v_{KS}(\boldsymbol{r})$, which also determines the exact ground-state density of the real molecular system of N interacting electrons moving in the external potential $v(\boldsymbol{r})$ due to the system nuclei in their fixed positions in space. The ground-state of the hypothetical non-interacting system is exactly described by a single KS determinant $\Psi^{KS}(N) \equiv |\psi| = \det\{\varphi_n\chi_n\}$ constructed from N orthonormal, *singly*-occupied spin-MO defined by the (real) spatial functions (MO) $\boldsymbol{\varphi} = \{\varphi_n(\boldsymbol{r})\}$ and the corresponding spin functions $\xi(\sigma) = \{\xi_n(\sigma)\}$, where the spin variable $\sigma = \pm\, ½$ [see Eq. (92)].

By the Hohenberg-Kohn theorem [41] both the effective potential $v_{KS}(\boldsymbol{r})$ and the system exact electronic energy are unique functionals of the ground–state electronic density, $v_{KS}(\boldsymbol{r}) = v_{KS}[\boldsymbol{r}; \rho]$ and $E[N, v] = E_v[\rho]$, where

$$\rho(\boldsymbol{r}) = \Sigma_n |\varphi_n(\boldsymbol{r})|^2 \equiv \Sigma_n \rho_n(\boldsymbol{r}). \tag{93}$$

The density functional for the ground-state electronic energy, $E_v[\rho]$, consists of the trivial external potential energy functional, $V_{ne}[\rho] = \int\rho(\boldsymbol{r})v(\boldsymbol{r})d\boldsymbol{r}$, and the universal ($v$-independent) Hohenberg-Kohn-Levy functional generating the sum of the expectation values of the electron kinetic and repulsion energies,

$$F[\rho] = T[\rho] + V_{ee}[\rho], \tag{94}$$

$$E_v[\rho] = \int\rho(\boldsymbol{r})\, v(\boldsymbol{r})\, d\boldsymbol{r} + F[\rho] \tag{95}$$

The total kinetic energy $T[\rho]$ of N electrons includes both the *non*-interacting (s) and Coulomb-correlation (c) parts:

$$T[\rho] = T_s[\rho] + T_c[\rho], \tag{96}$$

where

$$T_s[\rho] = -\frac{1}{2}\sum_{n=1}^{N}\langle\psi_n[\rho]|\Delta|\psi_n[\rho]\rangle = -\frac{1}{2}\sum_{n=1}^{N}\langle\varphi_n[\rho]|\Delta|\varphi_n[\rho]\rangle\,. \tag{97}$$

By the Hohenberg-Kohn variational principle the optimum KS orbitals, which also mark the solution point of the underlying KS-EPI principle, are the unique functionals of the system electron density: $\boldsymbol{\varphi} = \boldsymbol{\varphi}[\rho]$.

In KS theory one further extracts from the electron-repulsion energy of the real (interacting) system, $V_{ee}[\rho]$, the classical (Hartree) energy,

$$V_{ee}^{class}[\rho] = \frac{1}{2}\iint \frac{\rho(\boldsymbol{r})\rho(\boldsymbol{r}')}{|\boldsymbol{r}-\boldsymbol{r}'|}\, d\boldsymbol{r}\, d\boldsymbol{r}', \tag{98}$$

and the potential correlation contribution $E_{xc}[\rho] - T_c[\rho]$,

$$V_{ee}[\rho] = V_{ee}^{class}[\rho] + (E_{xc}[\rho] - T_c[\rho]), \tag{99}$$

so that

$$F[\rho] = T_s[\rho] + V_{ee}^{class}[\rho] + E_{xc}[\rho]. \tag{100}$$

Here $E_{xc}[\rho]$ stands for the *exchange-correlation* energy of KS theory, including the $T_c[\rho]$ contribution to $T[\rho]$, which determines the correlation part $v_{xc}(r)$ of KS potential,

$$v_{KS}(\boldsymbol{r}) = v(\boldsymbol{r}) + \int \frac{\rho(\boldsymbol{r}')}{|\boldsymbol{r}'-\boldsymbol{r}|}\, d\boldsymbol{r}' + \frac{\delta E_{xc}[\rho]}{\delta\rho(\boldsymbol{r})} \equiv v(\boldsymbol{r}) + v_H(\boldsymbol{r}) + v_{xc}(\boldsymbol{r}), \tag{101}$$

in the effective *one*-body Hamiltonian

$$\hat{\mathrm{H}}_{KS}(\boldsymbol{r}) = -\frac{1}{2}\Delta + v_{KS}(\boldsymbol{r}), \tag{102}$$

determining the optimum KS MO of N lowest eigenvalues (orbital energies) $\{\varepsilon_n\}$:

$$\hat{\mathrm{H}}_{KS}(\boldsymbol{r})\varphi_n(\boldsymbol{r}) = \varepsilon_n \varphi_n(\boldsymbol{r}), \qquad n = 1, 2, \ldots, N. \tag{103}$$

To summarize, in the KS theory the electronic energy of the interacting system is given by the following density functional:

$$E_v[\rho] = T_s[\rho] + V_{ne}[\rho] + V_{ee}^{class}[\rho] + E_{xc}[\rho] \equiv T_s[\varphi[\rho]] + V_e^{KS}[\rho], \qquad (104)$$

where the KS functional for the electronic potential energy $V_e^{KS}[\rho]$ contains the correlation part of the kinetic energy of interacting electrons:

$$V_e^{KS}[\rho] = (V_{ne}[\rho] + V_{ee}[\rho]) + T_c[\rho] \equiv V_e[\rho] + T_c[\rho]. \qquad (105)$$

The KS equations, which determine the optimum orbitals of the hypothetical non-interacting system, and hence also the electron density and energy of the interacting system, follow from the *Hohenberg-Kohn* (HK) variational principle for the system electronic energy,

$$\delta\{E_v[\rho] - \Sigma_n \Sigma_m \Theta_{mn} \langle\varphi_m|\varphi_n\rangle\} \equiv \delta K^{KS}[\varphi[\rho]] = 0 \qquad \text{or}$$

$$8\delta T_s[\varphi] \equiv \delta I^{KS}[\varphi[\rho]] = 8\Sigma_n\Sigma_m\Theta_{mn}\,\delta\langle\varphi_m|\varphi_n\rangle - 8\,V_e^{KS}[\rho] \equiv \delta J^{KS}[\varphi[\rho]], \qquad (106)$$

where, in the canonical representation the matrix of Lagrange multipliers, which enforce the MO normalization and orthogonality constraints, becomes diagonal, $\{\Theta_{mn} = \varepsilon_n\delta_{mn}\}$.

The preceding equation identifies the intrinsic ($I^{KS}[\varphi[\rho]]$) and bound ($J^{KS}[\varphi[\rho]]$) information terms in the KS EPI problem. Indeed, a straightforward integration by parts shows that the expectation value of the kinetic energy of non-interacting electrons of the separable KS system,

$$T_s[\varphi] = -\tfrac{1}{2}\,\Sigma_n \int \varphi_n(\boldsymbol{r})\nabla^2\varphi_n(\boldsymbol{r})\,d\boldsymbol{r} = \tfrac{1}{2}\,\Sigma_n \int [\nabla\varphi_n(\boldsymbol{r})]^2\,d\boldsymbol{r}, \qquad (107)$$

is proportional to the multi-component Fisher information $I[\psi] = I[\varphi]$:

$$\begin{aligned} I[\varphi] &= 4\,\Sigma_n \int [\nabla\varphi_n(\boldsymbol{r})]^2\,d\boldsymbol{r} \\ &= \Sigma_n \int [\nabla\rho_n(\boldsymbol{r})]^2/\rho_n(\boldsymbol{r})\,d\boldsymbol{r} = N\,\Sigma_n \int [\nabla p_n(\boldsymbol{r})]^2/p_n(\boldsymbol{r})\,d\boldsymbol{r} \\ &= 8T_s[\varphi] = I^{KS}[\varphi[\rho]]. \end{aligned} \qquad (108)$$

Thus, the KS variational rule of Eq.(106) can indeed be interpreted as another example of the EPI principle of Eq. (51):

$$\delta\{I^{KS}[\varphi[\rho]] - J^{KS}[\varphi[\rho]]\} = \delta K^{KS}[\varphi[\rho]] = 0. \tag{109}$$

In a similar way one derives the EPI problem for the orbital approximation in the wave-function theory of molecular electronic structure, which defines the familiar *Hartree-Fock* (HF) approach. Again, the kinetic energy functional of HF orbitals determines the intrinsic information part, while the combined electronic potential-energy terms supplemented by the constraints of the orbital orthonormality define the associated bound-information part of the EPI principle.

2.6. Information Principle for Adiabatic Separation of Electrons and Nuclei

As a final application of the physically-constrained Fisher-information principle let us examine the quantum mechanical system with particles differing in mass, in the familiar scenario of the BO separation of the (stationary) electronic and nuclear distributions in molecules. Here we shall focus on the Schrödinger equation for nuclear motions, in which the effect of the fast-moving (light) electrons is averaged out in the effective potential determining forces acting on the slowly-moving (heavy) nuclei. The Fisher information principle giving rise to the electronic Schrödinger equation, for the fixed (parametric) positions of nuclei, has already been discussed in Sections 2.3 and 2.5. The a.u. will again be used throughout this section.

A reference to Eq. (74) indicates that the Fisher information operator probes the length of the electron velocity, i.e., its momentum per unit particle mass. Therefore, in order to bring on equal footing the electron and nuclear Fisher intrinsic-information terms in the complete molecular EPI principle, which generates the adiabatic separation of the electronic and nuclear motions, one has to combine the electronic information term with the corresponding nuclear contributions per unit mass.

The molecular wave-function $\Psi(\mathbf{q},\mathbf{Q})$ of N electrons at positions $\mathbf{r} = \{\boldsymbol{r}_i\}$ exhibiting spins $\sigma = \{\sigma_i\}$, or in the combined notation $\mathbf{q} = \{\boldsymbol{r}_i, \sigma_i\} = \{\boldsymbol{q}_i\} = \{\mathbf{r}, \boldsymbol{\sigma}\}$ and m nuclei of masses $\{M_\alpha\}$, charges $\{Z_\alpha\}$, and spins $\boldsymbol{\Sigma} = \{\Sigma_\alpha\}$, in positions $\mathbf{R} = \{\boldsymbol{R}_\alpha\}$, $\mathbf{Q} = \{\boldsymbol{R}_\alpha, \Sigma_\alpha\} = \{\boldsymbol{Q}_\alpha\} = \{\mathbf{R}, \boldsymbol{\Sigma}\}$, generates the probability distribution of the joint electronic-nuclear events: $P(\mathbf{q},\mathbf{Q}) = |\Psi(\mathbf{q},\mathbf{Q})|^2$, satisfying the relevant overall and partial normalizations:

$$\int\int P(\mathbf{q},\mathbf{Q})\,d\mathbf{q}\,d\mathbf{Q} = \int \Pi(\mathbf{Q})\,d\mathbf{Q} = \int \pi(\mathbf{q})\,d\mathbf{q} = 1, \tag{110}$$

where $\Pi(\mathbf{Q})$ and $\pi(\mathbf{q})$ denote the partially-integrated nuclear and electronic probability distributions, respectively.

The essence of the adiabatic approximation lies in extracting the probability density of the heavy (slow) nuclei as the reference (parameter) distribution,

$$P(\mathbf{q},\mathbf{Q}) = \Pi(\mathbf{Q})\frac{P(\mathbf{q},\mathbf{Q})}{\Pi(\mathbf{Q})} \equiv \Pi(\mathbf{Q})\,p(\mathbf{q}|\mathbf{Q}),$$

$$\int p(\mathbf{q}|\mathbf{Q})\,d\mathbf{q} = 1, \tag{111}$$

where in the (conditional) electronic probability density $p(\mathbf{q}|\mathbf{Q})$ the nuclear variables $\mathbf{Q}$ constitute parameters, as indeed reflected by the normalization condition of this parametric probability density. This implies the familiar factorization of the system wave-function in terms of the nuclear and electronic functions $\chi(\mathbf{Q})$ and $\phi(\mathbf{q}|\mathbf{Q})$,

$$\Psi(\mathbf{q},\mathbf{Q}) = \chi(\mathbf{Q})\,\phi(\mathbf{q}|\mathbf{Q}), \tag{112}$$

representing the nuclear and conditional-electronic amplitudes of the associated probability distributions: $\Pi(\mathbf{Q}) = |\chi(\mathbf{Q})|^2$ and $p(\mathbf{q}|\mathbf{Q}) = |\phi(\mathbf{q}|\mathbf{Q})|^2$.

The intrinsic information functional for this molecular scenario of N electrons and m nuclei combines the average Fisher information terms per unit mass of all constituent particles:

$$I[\Psi] = I[P] = \langle I(N)\rangle + \sum_{\alpha=1}^{m}\frac{1}{M_\alpha}\langle I_\alpha\rangle. \tag{113}$$

Here the electronic contribution $\langle I(N)\rangle$ represents the nuclear-probability weighted mean-value of (equal) contributions due to the indistinguishable N electrons:

$$\langle I(N)\rangle = \sum_{i=1}^{N}\int\int\frac{[\nabla_i P(\mathbf{q},\mathbf{Q})]^2}{P(\mathbf{q},\mathbf{Q})}\,d\mathbf{q}\,d\mathbf{Q}$$

$$= 4\sum_{i=1}^{N} \int\int \nabla_i \Psi^* \cdot \nabla_i \Psi \, d\mathbf{q}\, d\mathbf{Q}$$

$$= 4\sum_{i=1}^{N} \int \Pi(\mathbf{Q})[\int \nabla_i \phi^* \cdot \nabla_i \phi \, d\mathbf{q}] d\mathbf{Q}$$

$$= \sum_{i=1}^{N} \int \Pi(\mathbf{Q}) \int \frac{[\nabla_i p(\mathbf{q}|\mathbf{Q})]^2}{p(\mathbf{q}|\mathbf{Q})} d\mathbf{q}\, d\mathbf{Q}$$

$$= \sum_{i=1}^{N} \int \Pi(\mathbf{Q}) \langle I_i(\mathbf{Q}) \rangle d\mathbf{Q} = N \int \Pi(\mathbf{Q}) \langle I_1(\mathbf{Q}) \rangle d\mathbf{Q}. \tag{114}$$

The average Fisher information due to the nucleus α similarly reads:

$$\langle I_\alpha \rangle = \int\int \frac{[\nabla_\alpha P(\mathbf{q},\mathbf{Q})]^2}{P(\mathbf{q},\mathbf{Q})} d\mathbf{q}\, d\mathbf{Q}$$

$$= \int \frac{[\nabla_\alpha \Pi(\mathbf{Q})]^2}{\Pi(\mathbf{Q})} d\mathbf{Q} + \int \Pi(\mathbf{Q}) \int \frac{[\nabla_\alpha p(\mathbf{q}|\mathbf{Q})]^2}{p(\mathbf{q}|\mathbf{Q})} d\mathbf{q}\, d\mathbf{Q}$$

$$+ \int \nabla_\alpha \Pi(\mathbf{Q}) \cdot \int \nabla_\alpha p(\mathbf{q}|\mathbf{Q}) d\mathbf{q}\, d\mathbf{Q}$$

$$= 4\int\int \nabla_\alpha \Psi^* \cdot \nabla_\alpha \Psi \, d\mathbf{q}\, d\mathbf{Q}$$

$$= 4\Big(\int |\nabla_\alpha \chi|^2 d\mathbf{Q} + \int \Pi \int |\nabla_\alpha \phi|^2 d\mathbf{q}\, d\mathbf{Q} + \int \chi(\nabla_\alpha \chi^*) \cdot \int \phi^* \nabla_\alpha \phi \, d\mathbf{q}\, d\mathbf{Q}$$

$$+ \int \chi^*(\nabla_\alpha \chi) \cdot \int \phi \nabla_\alpha \phi^* d\mathbf{q}\, d\mathbf{Q} \Big). \tag{115}$$

The molecular EPI principle should now include, besides the usual subsidiary conditions of the probability normalizations, the constraint of the fixed value of the overall Coulombic potential energy:

$$\langle V \rangle = \int\int P(\mathbf{q},\mathbf{Q}) V(\mathbf{q},\mathbf{Q}) \, d\mathbf{q}\, d\mathbf{Q}$$

$$= \int \Pi(\mathbf{Q}) \int p(\mathbf{q}|\mathbf{Q})[V_{ne}(\mathbf{q},\mathbf{Q}) + V_{ee}(\mathbf{q})]\, d\mathbf{q}\, d\mathbf{Q} + \int \Pi(\mathbf{Q}) V_{nn}(\mathbf{Q})\, d\mathbf{Q}$$

$$\equiv \int \Pi(\mathbf{Q}) \langle V(\mathbf{Q})\rangle_{\mathbf{q}}\, d\mathbf{Q} = \langle V \rangle^{0}, \tag{116}$$

where $\langle V(\mathbf{Q})\rangle_{\mathbf{q}}$ denotes the overall potential energy for the specified nuclear positions, averaged over the electronic coordinates $\mathbf{q}$, and the multiplicative operator

$$V(\mathbf{q},\mathbf{Q}) = -\sum_{\alpha=1}^{m}\sum_{i=1}^{N} \frac{Z_\alpha}{|\boldsymbol{r}_i - \boldsymbol{R}_\alpha|} + \sum_{i=1}^{N-1}\sum_{j=i+1}^{N} \frac{1}{|\boldsymbol{r}_i - \boldsymbol{r}_j|} + \sum_{\alpha=1}^{m-1}\sum_{\beta=\alpha+1}^{m} \frac{Z_\alpha Z_\beta}{|\boldsymbol{R}_\alpha - \boldsymbol{R}_\beta|}$$

$$\equiv V_{ne}(\mathbf{q},\mathbf{Q}) + V_{ee}(\mathbf{q}) + V_{nn}(\mathbf{Q})\,. \tag{117}$$

The resulting adiabatic Fisher information principle

$$\delta\{I[P] - \lambda_1 \int [\Pi(\mathbf{Q}) - \lambda_2(\mathbf{Q}) \int p(\mathbf{q}\,|\,\mathbf{Q})\, d\mathbf{q}]\, d\mathbf{Q}$$

$$- \lambda_3 \int \Pi(\mathbf{Q}) \int p(\mathbf{q}\,|\,\mathbf{Q}) V(\mathbf{q},\mathbf{Q})\, d\mathbf{q}\, d\mathbf{Q}\} = 0, \tag{118}$$

includes the global (λ_1,λ_3) and local [$\lambda_2(\mathbf{Q})$] Lagrange multipliers, which respectively enforce the subsidiary conditions of the normalization of $\Pi(\mathbf{Q})$, the fixed value of the average potential energy, and the normalization of $p(\mathbf{q}|\mathbf{Q})$ for any given $\mathbf{Q}$.

The functional differentiation of the overall Fisher information of Eq. (113) with respect to $\chi^*(\mathbf{Q})$ then gives:

$$\frac{\delta I[\Psi]}{\delta \chi^*(\mathbf{Q})} = 4\Bigg\{\left(\sum_{i=1}^{N} \int |\nabla_i \phi|^2 d\mathbf{q}\right)\chi(\mathbf{Q})$$

$$-\sum_{\alpha=1}^{m} \frac{1}{M_\alpha}\Big[\Delta_\alpha \chi(\mathbf{Q}) + \left(\int \phi^* \Delta_\alpha \phi\, d\mathbf{q}\right)\chi(\mathbf{Q}) + \left(\int \phi^* \nabla_\alpha \phi\, d\mathbf{q}\right)\cdot \nabla_\alpha \chi(\mathbf{Q})\Big]\Bigg\} \tag{119}$$

where we have taken into account the conservation of the normalization of the electronic (conditional) probability, which further implies

$$\nabla_\alpha \int p(\mathbf{q}|\mathbf{Q})\,d\mathbf{q} = \int (\phi^* \nabla_\alpha \phi + \phi \nabla_\alpha \phi^*)\; d\mathbf{q} = 0\,. \tag{120}$$

The differentiation of the constraint part of the auxiliary functional of Eq. (118) generates the remaining terms of the Euler equation for the optimum nuclear wave-function of Eq. (112):

$$\frac{\delta I[\Psi]}{\delta \chi^*(\mathbf{Q})} - [\lambda_1 + \lambda_3 \langle V(\mathbf{Q})\rangle_{\mathbf{q}}]\chi(\mathbf{Q}) = 0\,. \tag{121}$$

Next, let us interpret the contributions in Eq. (119) in terms of the electronic and nuclear kinetic-energy operators:

$$\hat{\mathrm{T}}(\mathbf{q}) = -1/2\sum_{i=1}^{N}\Delta_i \qquad \text{and} \qquad \hat{\mathrm{T}}_n(\mathbf{Q}) = -\sum_{\alpha=1}^{m}\frac{1}{2M_\alpha}\Delta_\alpha\,. \tag{122}$$

The nuclear Euler equation then reads:

$$\left\{\hat{\mathrm{T}}_n(\mathbf{Q}) + [\langle T(\mathbf{Q})\rangle_{\mathbf{q}} - \frac{\lambda_3}{8}\langle V(\mathbf{Q})\rangle_{\mathbf{q}}] + \langle T_n^\phi(\mathbf{Q})\rangle_{\mathbf{q}} - \frac{\lambda_1}{8}\right\}\chi(\mathbf{Q})$$

$$-\sum_{\alpha=1}^{m}\frac{1}{2M_\alpha}\langle\phi|\nabla_\alpha\phi\rangle_{\mathbf{q}}\cdot\nabla_\alpha\chi(\mathbf{Q}) = 0\,, \tag{123}$$

where

$$\langle T_n^\phi(\mathbf{Q})\rangle_{\mathbf{q}} = \langle\phi|\hat{\mathrm{T}}_n|\phi\rangle_{\mathbf{q}} \qquad\qquad 1 \tag{124}$$

stands for the ("diagonal") kinetic-energy correction to the electronic *Potential Energy Surface* (PES)

$$\langle E(\mathbf{Q})\rangle_{\mathrm{q}} = \langle T(\mathbf{Q})\rangle_{\mathbf{q}} + \langle V(\mathbf{Q})\rangle_{\mathbf{q}} = E_e(\mathbf{Q}) \tag{125}$$

in the resultant effective potential for nuclear motions in the familiar adiabatic approximation,

$$U(\mathbf{Q}) = E_e(\mathbf{Q}) + \left\langle T_n^{\phi}(\mathbf{Q})\right\rangle_{\mathbf{q}}. \tag{126}$$

It complements $\hat{\mathrm{T}}_n(\mathbf{Q})$ in the effective Hamiltonian

$$\hat{\mathrm{H}}_n^{eff.}(\mathbf{Q}) = \hat{\mathrm{T}}_n(\mathbf{Q}) + U(\mathbf{Q}),$$

which accounts for the averaged-out (integrated) influence due to fast electronic motions, of the Schrödinger equation determining the nuclear distributions:

$$\{\hat{\mathrm{T}}_n(\mathbf{Q}) + U(\mathbf{Q}) - E_{mol.}\}\chi(\mathbf{Q}) \equiv [\hat{\mathrm{H}}_n^{eff.}(\mathbf{Q}) - E_{mol.}]\chi(\mathbf{Q}) = 0; \tag{127}$$

here $E_{mol.}$ denotes the full Coulomb molecular energy, the eigenvalue of the effective Hamiltonian $\hat{\mathrm{H}}_n^{eff.}(\mathbf{Q})$.

In this approximation one neglects the last, *non-adiabatic* term in Eq. (123), which involves the nuclear gradients of both the electronic and nuclear wave-functions. Equation (123) is seen to assume the form of the preceding eigenvalue equation of the effective nuclear Hamiltonian for $\lambda_3 = -8$ and $\lambda_1 = 8\ E_{mol.}$. It should be also recalled that by additionally neglecting the small diagonal correction $\left\langle T_n^{\phi}(\mathbf{Q})\right\rangle_{\mathbf{q}}$ in the effective potential for nuclear motions, $U(\mathbf{Q}) \approx E_e(\mathbf{Q})$, one arrives at the original (crude adiabatic) approximation of Born and Oppenheimer.

2.7. Superposition Principles

For reasons of simplicity, let us again consider the quantum mechanical mixing of two (complex) orthonormal states of a single spin-less particle of mass μ, $\psi = \{\psi_k = R_k\exp(i\Phi_k\}, \langle\psi_k | \psi_l\rangle = \delta_{k,l}$; $k, l = 1, 2$, where the probability amplitude R_k and the phase Φ_k parts in general depend upon the position coordinates and time [Eq. (53)]:

$$\psi = 2^{-1/2}(\psi_1 + \psi_2) \equiv C_1\psi_1 + C_2\psi_2. \tag{128}$$

Expressing the total probability density, $p(\boldsymbol{r}) = |\psi(\boldsymbol{r})|^2$, in terms of probabilities $\{p_k(\boldsymbol{r}) = |\psi_k(\boldsymbol{r})|^2 = \langle\psi_k|\boldsymbol{r}\rangle\langle\boldsymbol{r}|\psi_k\rangle \equiv \langle\psi_k|\hat{\mathrm{p}}(\boldsymbol{r})|\psi_k\rangle\}$ and phases of individual states in this combination then gives the familiar result:

$$p = (|C_1|^2\,|\psi_1|^2 + |C_1|^2\,|\psi_1|^2) + (C_1^* C_2 \psi_1^* \psi_2 + C_2^* C_1 \psi_2^* \psi_1)$$

$$= \frac{1}{2}(p_1 + p_2) + \sqrt{p_1 p_2}\,\cos(\Phi_1 - \Phi_2) \equiv p_\psi^{add.} + p_\psi^{nadd.}. \qquad (129)$$

It identifies the third, *superposition* term, depending upon the relative phases of two functions in the combined state, as the *non-additive* probability contribution

$$p_\psi^{nadd.} = p_\psi^{total} - p_\psi^{add.}, \qquad (130)$$

expressed as the difference between the *total* probability in the ψ-resolution,

$$p_\psi^{total} = p = |\psi|^2 = ½\,|\psi_1 + \psi_2|^2, \qquad (131)$$

and its additive contribution given by the weighted average of the probability distributions of the combined states,

$$p_\psi^{add.} = P(\psi_1|\psi)\,p_1 + P(\psi_2|\psi)\,p_2, \qquad P(\psi_1|\psi) + P(\psi_2|\psi) = 1. \qquad (132)$$

Indeed, in accordance with the quantum-mechanical superposition principle [45] the weights provided by the squares of the combination coefficients $\{C_i\}$,

$$\{P(\psi_i|\psi) = |\langle\psi|\psi_i\rangle|^2\} = |C_i|^2 = ½\}$$

define the relevant *conditional* probabilities $P(\psi_1|\psi) = P(\psi_2|\psi) = ½$ of individual states in the ensemble defined by the density operator

$$\hat{D}_\psi = P(\psi_1|\psi)\,|\psi_1\rangle\langle\psi_1| + P(\psi_2|\psi)\,|\psi_2\rangle\langle\psi_2| = \frac{1}{2}\left(|\psi_1\rangle\langle\psi_1| + |\psi_2\rangle\langle\psi_2|\right), \qquad (133)$$

in terms of which

$$p_\psi^{add.}(\boldsymbol{r}) = \mathrm{Tr}[\hat{D}_\psi\,\hat{p}(\boldsymbol{r})] = 1/2[p_1(\boldsymbol{r}) + p_2(\boldsymbol{r})].$$

The probability-interference term,

$$p_{\psi}^{nadd.} = \sqrt{p_1 p_2}\cos(\Phi_1 - \Phi_2), \tag{134}$$

is responsible for the chemical bond, say, between two hydrogen atoms, when the two AO's of constituent atoms are mixed into the symmetric (doubly occupied) bonding MO [7,24]. Clearly, in the real AO case, when $\Phi_1 = \Phi_2 = 0$, this contribution reduces into the geometric average of the component probability densities: $p_{\psi}^{nadd.} = \sqrt{p_1 p_2}$. It should be also observed that the overall probability p determines the modulus factor of $\psi = R\exp(i\Phi)$, $R = \sqrt{p}$, while its resultant phase follows from the equation $\Phi = \arccos[R^{-1}\mathrm{Re}(\psi)] = \arcsin[R^{-1}\mathrm{Im}(\psi)]$.

Next, let us examine the related superposition rule for the density $f(\boldsymbol{r})$ of the generalized Fisher-information in the combined state ψ,

$$f(\boldsymbol{r}) \equiv f_{\psi}^{total}(\boldsymbol{r}) = f_{\psi}^{add.}(\boldsymbol{r}) + f_{\psi}^{nadd.}(\boldsymbol{r}), \tag{135}$$

measuring the modulus of the gradient content of this (complex) probability amplitude:

$$f \equiv f[\psi] = 4\nabla\psi^* \cdot \nabla\psi = 4[(\nabla R)^2 + R^2(\nabla\Phi)^2]$$

$$= (\nabla p)^2/p + \frac{4\mu^2}{\hbar^2}\frac{\boldsymbol{j}^2}{p} \equiv f[p] + f[\boldsymbol{j}]. \tag{136}$$

As indicated in Eq. (135) this information density defines the total contribution in the ψ-resolution, $f_{\psi}^{total} \equiv f[\psi]$, while the weighted sum of the information content of two individual states $\{f[\psi_k] = 4\nabla\psi_k^* \cdot \nabla\psi_k \equiv f_k\}$ determines its additive component,

$$f_{\psi}^{add.} = \mathrm{Tr}[\hat{\mathrm{D}}_{\psi}\hat{\mathrm{I}}(\boldsymbol{r})] = P(\psi_1|\psi)\, f[\psi_1] + P(\psi_2|\psi)\, f[\psi_2] \equiv ½\,(f_1 + f_2). \tag{137}$$

Here

$$f_k = 4[(\nabla R_k)^2 + R_k^2(\nabla\Phi_k)^2] = (\nabla p_k)^2/p_k + \frac{4\mu^2}{\hbar^2}\frac{\boldsymbol{j}_k^2}{p_k} \equiv f_k[p_k] + f_k[\boldsymbol{j}_k], \tag{138}$$

where $\boldsymbol{j}_k \equiv \boldsymbol{j}[\psi_k]$ denotes the probability-current density in the member state ψ_k:

$$\boldsymbol{j}_k = \frac{\hbar}{2\mu i}(\psi_k^*\nabla\psi_k - \psi_k\nabla\psi_k^*) = \frac{\hbar}{\mu}\mathrm{Im}(\psi_k^*\nabla\psi_k) = \frac{\hbar}{\mu}R_k^2\nabla\Phi_k = P_k\nabla\left[\frac{\hbar\Phi_k}{\mu}\right]. \quad (139)$$

The resulting non-additive (interference) information density in terms of the (real) probability amplitudes and phases of the two combined states now reads:

$$f_\psi^{nadd.} = 4\,[C_1^*C_2\nabla\psi_1^*\cdot\nabla\psi_2 + C_2^*C_1\nabla\psi_2^*\cdot\nabla\psi_1]$$

$$= 2\,[\nabla\psi_1^*\cdot\nabla\psi_2 + \nabla\psi_2^*\cdot\nabla\psi_1]$$

$$= 4\,[(\nabla R_1\cdot\nabla R_2 + R_1R_2\nabla\Phi_1\cdot\nabla\Phi_2)\cos(\Phi_1 - \Phi_2)$$

$$+ (R_2\nabla R_1\cdot\nabla\Phi_2 - R_1\nabla R_2\cdot\nabla\Phi_1)\sin(\Phi_1 - \Phi_2)]. \quad (140)$$

This expression can be somewhat simplified, when expressed in terms of the probability and current descriptors $\{p_k, \boldsymbol{j}_k\}$ of the two individual states. Eliminating $\nabla\psi_k$ from the relevant expressions for $\nabla p_k = \psi_k\nabla\psi_k^* + \psi_k^*\nabla\psi_k$ and $\boldsymbol{j}_k$ gives:

$$\nabla\psi_k = \frac{\psi_k}{2p_k}\left(\nabla p_k + \frac{2\mu i}{\hbar}\boldsymbol{j}_k\right).$$

Substituting the preceding equation and its complex conjugate into Eq. (140) then gives the following expression for the non-additive term in the information superposition principle of Eq. (135):

$$f_\psi^{nadd.} = \frac{1}{\sqrt{p_1p_2}}\left[\left(\nabla p_1\cdot\nabla p_2 + \frac{4\mu^2}{\hbar^2}\boldsymbol{j}_1\cdot\boldsymbol{j}_2\right)\cos(\Phi_1 - \Phi_2) + \frac{2\mu}{\hbar}(\nabla p_1\cdot\boldsymbol{j}_2 - \nabla p_2\cdot\boldsymbol{j}_1)\sin(\Phi_1 - \Phi_2)\right]$$

$$\equiv (\overline{\nabla}p_1\cdot\overline{\nabla}p_2 + \bar{\boldsymbol{j}}_1\cdot\bar{\boldsymbol{j}}_2)\cos(\Phi_1 - \Phi_2) + (\overline{\nabla}p_1\cdot\bar{\boldsymbol{j}}_2 - \overline{\nabla}p_2\cdot\bar{\boldsymbol{j}}_1)\sin(\Phi_1 - \Phi_2), \quad (141)$$

where bars denote the reduced quantities: $\overline{\nabla}p_k = p_k^{-1/2}\nabla p_k$ and $\bar{\boldsymbol{j}}_k = 2\mu/(\hbar p_k^{1/2})\boldsymbol{j}_k$. The preceding equation expresses the change in the Fisher information density, relative to the reference level of the additive contribution of Eq. (137), which accompanies the quantum-mechanical superposition of two individual states. For the stationary or real member states, when the current-dependent contributions identically vanish, e.g., in combining two real AO into MO, this non-additive information contribution is thus seen to be solely determined by the product of the reduced gradients of the particle probability distributions in the combined states.

Of interest also is the partitioning of the probability-current density of the combination of two states, into the corresponding additive and non-additive contributions:

$$\boldsymbol{j} = \boldsymbol{j}[\psi] \equiv \boldsymbol{j}_{\psi}^{total} = \boldsymbol{j}_{\psi}^{add.} + \boldsymbol{j}_{\psi}^{nadd.} . \tag{142a}$$

Here,

$$\boldsymbol{j}_{\psi}^{add.} = \mathrm{Tr}[\hat{\mathrm{D}}_{\psi}\hat{\mathrm{j}}(\boldsymbol{r})] = P(\psi_1|\psi)\,\boldsymbol{j}_1 + P(\psi_2|\psi)\,\boldsymbol{j}_2 = (\boldsymbol{j}_1 + \boldsymbol{j}_2)/2, \tag{142b}$$

where $\hat{\mathrm{j}}(\boldsymbol{r})$ denotes the current density operator (see Appendix A), and

$$\boldsymbol{j}_{\psi}^{nadd.} = \frac{\hbar}{2\mu i}\,[C_1^* C_2[\psi_1^* \nabla \psi_2 - \psi_2 \nabla \psi_1^*) + C_2^* C_1(\psi_2^* \nabla \psi_1 - \psi_1 \nabla \psi_2^*)]$$

$$= \frac{\hbar}{4\mu i}\,[\psi_1^* \nabla \psi_2 - \psi_1 \nabla \psi_2^* + \psi_2^* \nabla \psi_1 - \psi_2 \nabla \psi_1^*]$$

$$= \frac{\hbar}{4\mu}\,[(\sqrt{p_2}\,\bar{\nabla} p_1 - \sqrt{p_1}\,\bar{\nabla} p_2)\,\sin(\Phi_1 - \Phi_2)$$

$$+ (\sqrt{p_2}\,\bar{\boldsymbol{j}}_1 + \sqrt{p_1}\,\bar{\boldsymbol{j}}_2)\,\cos(\Phi_1 - \Phi_2)] . \tag{142c}$$

Therefore, the interference contribution to the stream of probability density again depends on the relative phases of the two combined states. For the real member functions, for which the relative phase and currents identically vanish, the above additive and non-additive flows of probability vanish identically since then $\boldsymbol{j}_1 = \boldsymbol{j}_2 = \boldsymbol{0}$ and $\Phi_1 - \Phi_2 = 0$.

Chapter 3

ELECTRON DISTRIBUTIONS AS CARRIERS OF INFORMATION IN MOLECULES

3.1. ALTERNATIVE LOCAL INFORMATION PROBES OF CHEMICAL BONDS

The electron densities $\{\rho_i^0\}$ of the separated atoms define the molecular (isoelectronic, $N = N^0$) prototype called the atomic "promolecule" [7,23], given by the sum of the free-atom electron densities shifted to the actual locations of atoms in the molecule. The resulting electron density $\rho^0 = \sum_i \rho_i^0$ of this collection of the "frozen" free-atom distributions defines the initial stage in the bond-formation process and determines a natural reference for extracting changes due to the chemical bonds in the *density difference* function $\Delta\rho = \rho - \rho^0$. This deformation density has been widely used to probe the electronic structure of molecular systems. In this section we shall use some local IT probes to explore the molecular ground-state electron distributions $\rho(\boldsymbol{r})$ or their shape (probability)-factors $p(\boldsymbol{r})=\rho(\boldsymbol{r})/N$, generated by the KS calculations in the *Local Density Approximation* (LDA) of DFT.

Consider first the density $\Delta s(\boldsymbol{r})$ (in nats per unit volume) of the KL entropy-deficiency (directed-divergence) functional, reflecting the information distance between the molecular and promolecular electron distributions:

$$\Delta S[\rho|\rho^0] = \int \rho(\boldsymbol{r}) \ln[\rho(\boldsymbol{r})/\rho^0(\boldsymbol{r})]\, d\boldsymbol{r} = \int \rho(\boldsymbol{r})\, I[w(\boldsymbol{r})]\, d\boldsymbol{r} \equiv \int \Delta s(\boldsymbol{r})\, d\boldsymbol{r}, \qquad (143)$$

where $w(\boldsymbol{r}) = \rho(\boldsymbol{r})/\rho^0(\boldsymbol{r}) = p(\boldsymbol{r})/p^0(\boldsymbol{r})$ stands for the local density/probability enhancement factor and $I[w(\boldsymbol{r})]$ denotes the *surprisal* function. This functional also represents the (renormalized) missing information between the shape(probability)-factors $p(\boldsymbol{r})$ and $p^0(\boldsymbol{r})=\rho^0(\boldsymbol{r})/N^0$ of the two compared electron densities:

$$\Delta s(\boldsymbol{r}) \equiv \Delta s[\rho(\boldsymbol{r})|\rho^0(\boldsymbol{r})] = Np(\boldsymbol{r}) \ln[p(\boldsymbol{r})/p^0(\boldsymbol{r})] \equiv N\,\Delta s[p(\boldsymbol{r})|p^0(\boldsymbol{r})]. \tag{144}$$

This KL-information density measures the average, electron density/ probability weighted surprisal. The related divergence measure of Kullback's missing information reads:

$$\Delta S[\rho, \rho^0] = \int[\rho(\boldsymbol{r}) - \rho^0(\boldsymbol{r})] \ln[\rho(\boldsymbol{r})/\rho^0(\boldsymbol{r})]\, d\boldsymbol{r} \equiv \int\Delta\rho(\boldsymbol{r})\, I[w(\boldsymbol{r})]\, d\boldsymbol{r} \equiv \int\Delta D(\boldsymbol{r})\, d\boldsymbol{r}. \tag{145}$$

Its density $\Delta D(\boldsymbol{r})$ represents the $\Delta\rho(\boldsymbol{r}) = \rho(\boldsymbol{r}) - \rho^0(\boldsymbol{r})$ (or $\Delta p(\boldsymbol{r}) = p(\boldsymbol{r}) - p^0(\boldsymbol{r}) = \Delta\rho(\boldsymbol{r})/N$) weighted surprisal:

$$\Delta D(\boldsymbol{r}) \equiv \Delta s[\rho(\boldsymbol{r}), \rho^0(\boldsymbol{r})] = \Delta\rho(\boldsymbol{r})\, I[w(\boldsymbol{r})] = N\Delta p(\boldsymbol{r})\, I[w(\boldsymbol{r})] = N\,\Delta s[p(\boldsymbol{r}), p^0(\boldsymbol{r})]. \tag{146}$$

Therefore, the molecular surprisal $I[w(\boldsymbol{r})]$ itself measures the density-per-electron of the local entropy-deficiency relative to the promolecular reference,

$$I[w(\boldsymbol{r})] = \Delta s(\boldsymbol{r})/\rho(\boldsymbol{r}), \tag{147}$$

or the density-per-electron-displacement of the molecular divergence,

$$I[w(\boldsymbol{r})] = \Delta D(\boldsymbol{r})/\Delta\rho(\boldsymbol{r}). \tag{148}$$

We shall also examine the molecular displacements in the average Shannon entropy, relative to the promolecular reference value,

$$\mathscr{H}[\rho] \equiv S[\rho] - S[\rho^0] = -\int\rho(\boldsymbol{r}) \ln\rho(\boldsymbol{r})\, d\boldsymbol{r} + \int\rho^0(\boldsymbol{r}) \ln\rho^0(\boldsymbol{r})\, d\boldsymbol{r} \equiv \int h_\rho(\boldsymbol{r})\, d\boldsymbol{r}, \tag{149}$$

and its density $h_\rho(\boldsymbol{r})$ as alternative global and local probes of electron distributions in molecules. Alternatively, the corresponding entropy shifts in terms of the (normalized) probability distributions,

$$\mathscr{H}[p] \equiv S[p] - S[p^0] = -\int p(\boldsymbol{r}) \ln p(\boldsymbol{r})\, d\boldsymbol{r} + \int p^0(\boldsymbol{r}) \ln p^0(\boldsymbol{r})\, d\boldsymbol{r} \equiv \int h_p(\boldsymbol{r})\, d\boldsymbol{r}, \tag{150}$$

has been used to explore the local relaxation in electron uncertainties, which accompanies the bond formation in molecules.

In fact all these information densities can be regarded as complementary tools for diagnosing the presence of chemical bonds and for monitoring the effective valence states of the bonded atoms in the molecular environment.

We further observe that the molecular electron density $\rho(\boldsymbol{r})$ is on average only slightly modified relative to the promolecular distribution $\rho^0(\boldsymbol{r})$, $\rho(\boldsymbol{r}) \approx \rho^0(\boldsymbol{r})$ or $w(\boldsymbol{r}) \approx 1$. Indeed, the formation of chemical bonds involves only a minor reconstruction of the electronic structure, mainly in the *valence*-shells of the constituent atoms, so that $|\Delta\rho(\boldsymbol{r})| \equiv |\rho(\boldsymbol{r}) - \rho^0(\boldsymbol{r})| << \rho(\boldsymbol{r}) \approx \rho^0(\boldsymbol{r})$ and hence the ratio $\Delta\rho(\boldsymbol{r})/\rho(\boldsymbol{r}) \approx \Delta\rho(\boldsymbol{r})/\rho^0(\boldsymbol{r})$ is generally small in the energetically important regions of large density values. As explicitly shown in the first column of Figure 2, the largest values of the density difference $\Delta\rho(\boldsymbol{r})$ are observed mainly in the bond region, between the nuclei of chemically bonded atoms; the reconstruction of atomic lone pairs can also lead to an appreciable displacement in the molecular electron density.

By expanding the logarithm of the molecular surprisal $I[w(\boldsymbol{r})]$ around $w(\boldsymbol{r}) = 1$ to *first*-order in the relative displacement of the electron density one obtains the following approximate relations between the local value of the molecular surprisal and that of the density-difference function:

$$I[w(\boldsymbol{r})] = \ln[\rho(\boldsymbol{r})/\rho^0(\boldsymbol{r})] = \ln\{[\rho^0(\boldsymbol{r}) + \Delta\rho(\boldsymbol{r})]/\rho^0(\boldsymbol{r})\}$$

$$\cong \Delta\rho(\boldsymbol{r})/\rho^0(\boldsymbol{r}) \approx \Delta\rho(\boldsymbol{r})/\rho(\boldsymbol{r}). \tag{151}$$

This relation provides a semi-quantitative information-theoretic interpretation of the relative density difference diagrams and links the local surprisal of IT to the density difference function of quantum chemistry [7,9,10]. It also relates the integrands of alternative information-distance functionals to the corresponding functions of displacements in the electron density:

$$\Delta s(\boldsymbol{r}) = \rho(\boldsymbol{r})I[w(\boldsymbol{r})] \cong \Delta\rho(\boldsymbol{r})\, w(\boldsymbol{r}) \approx \Delta\rho(\boldsymbol{r}), \tag{152}$$

$$\Delta D(\boldsymbol{r}) = \Delta\rho(\boldsymbol{r})\; I[w(\boldsymbol{r})] \cong [\Delta\rho(\boldsymbol{r})]^2/\rho^0(\boldsymbol{r}) \geq 0. \tag{153}$$

The first of these relations qualitatively explains a remarkable similarity between the density difference and the KL information density plots observed in Figure 2, where the contour maps for selected linear diatomics and triatomics are reported.

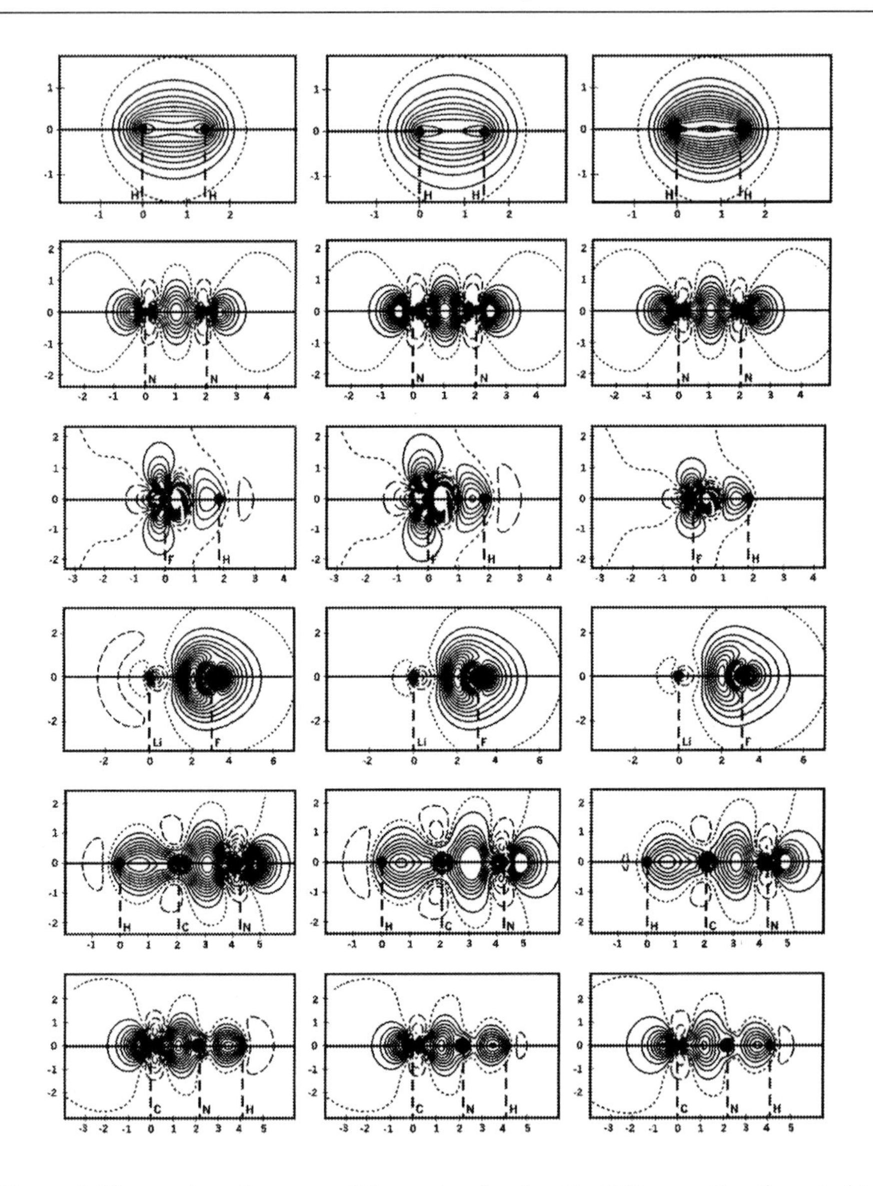

Figure 2. The contour diagrams of the molecular density difference function, $\Delta\rho(\boldsymbol{r}) = \rho(\boldsymbol{r}) - \rho^0(\boldsymbol{r})$ (first column), the information-distance density, $\Delta s(\boldsymbol{r}) = \rho(\boldsymbol{r})I[w(\boldsymbol{r})]$ (second column) and its approximate, *first*-order expansion, $\Delta s(\boldsymbol{r}) \cong \Delta\rho(\boldsymbol{r})w(\boldsymbol{r})$ (third column), for selected diatomic and linear triatomic molecules: H_2, HF, LiF, HCN and HNC. The solid, pointed and broken lines denote the positive, zero and negative values, respectively, of the equally spaced contours [7,9]; the same convention is applied in Figures 3, 4, 6, 10, and 11

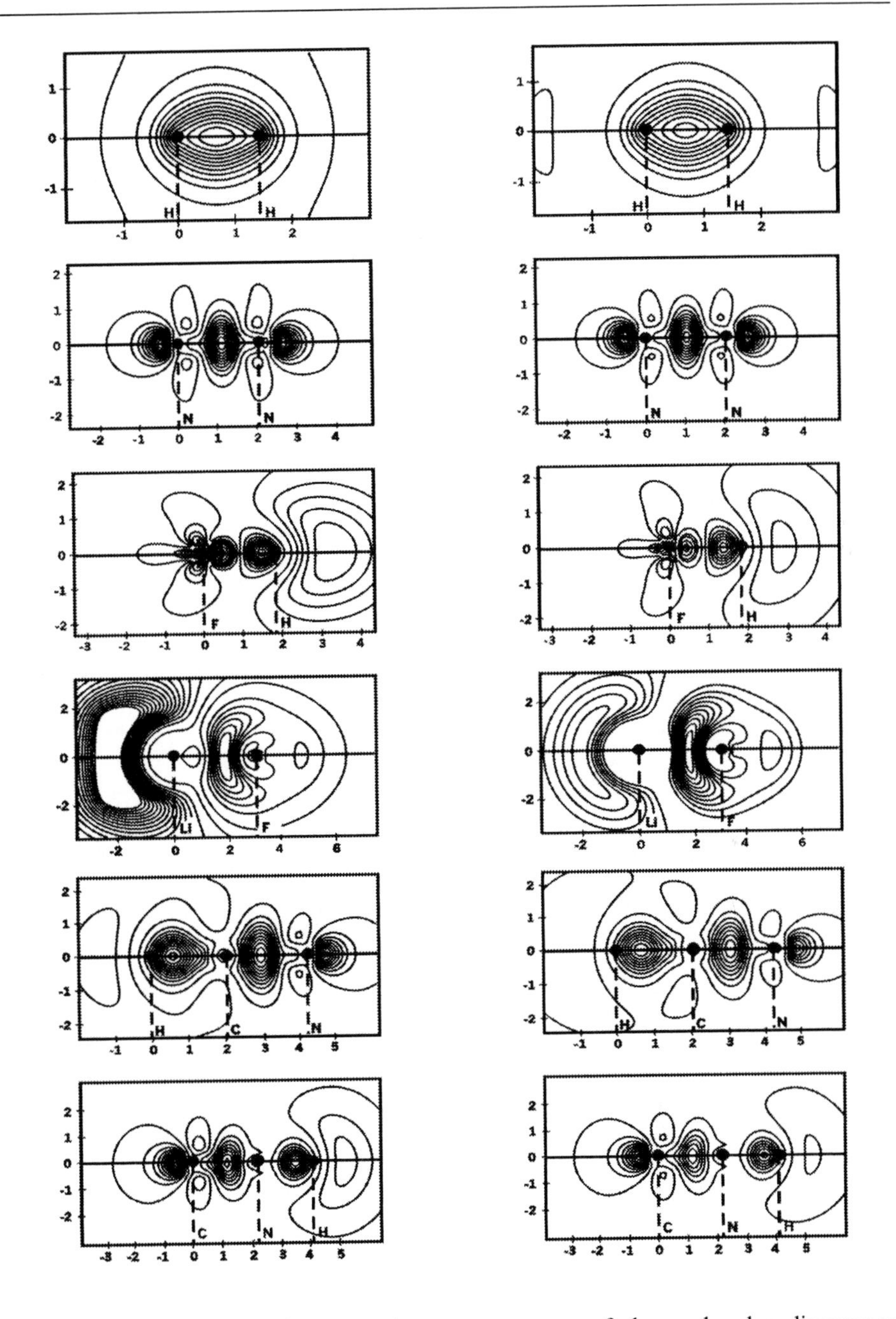

Figure 3. A comparison between the contour maps of the molecular divergence density, $\Delta D(\boldsymbol{r}) = \Delta\rho(\boldsymbol{r})I[w(\boldsymbol{r})]$ (first column), $[\Delta\rho(\boldsymbol{r})]^2/\rho^0(\boldsymbol{r})$ (second column), for the molecules of Figure 2, which validates the approximate relation of Eq. (55) [7,9]

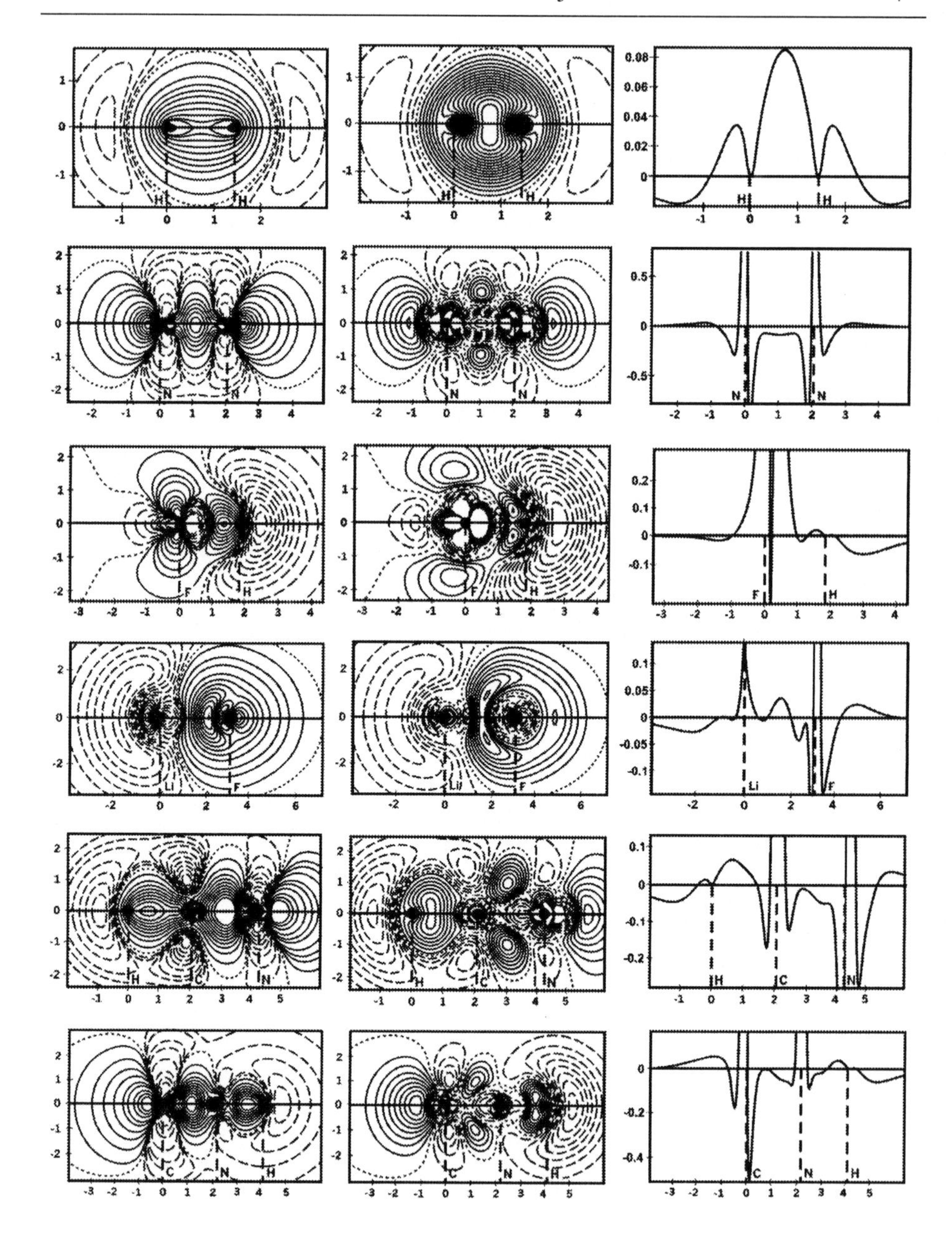

Figure 4. A comparison between the (non-equidistant) contour diagrams of the density difference $\Delta\rho(\boldsymbol{r})$ (first column) and entropy-difference $h_\rho(\boldsymbol{r})$ (second column) functions for the linear molecules of Figure 2. The corresponding profiles of $h_\rho(\boldsymbol{r})$ for the cuts along the bond axis are shown in the third column of the figure [7,10]

3.2. Comparison between Density-Difference and Information-Distance Diagrams

The preceding approximate equalities are numerically verified in Figures 2 and 3, respectively. In the former, the contour diagram of the directed divergence density $\Delta s(\boldsymbol{r})$ (second column) is compared with the corresponding map of its *first*-order approximation [Eq. (152)], $\Delta\rho(\boldsymbol{r})w(\boldsymbol{r})$ (third column), and the density difference itself (first column). A general similarity between these diagrams in each row of the figure confirms a semi-quantitative character of the *first*-order expansions of the directed divergence densities. The corresponding numerical validation of Eq. (153) is shown in Figure 3, where the contour maps of Kullback's divergence density $\Delta D(\boldsymbol{r})$ (first column) are compared with the corresponding diagrams of its first-order approximation $[\Delta\rho(\boldsymbol{r})]^2/\rho^0(\boldsymbol{r})$ (second column). Again, a semi-quantitative similarity between the two diagrams in each row numerically validates this approximate relation.

In Figure 2 the density difference function $\Delta\rho(\boldsymbol{r})$ for representative linear diatomic and triatomic molecules exhibits typical aspects of the equilibrium reconstructions of the free-atoms during formation of the single and multiple chemical bonds, which exhibit varying degree of the bond covalency (electron-sharing) and ionicity (electron transfer) components.

Let us first examine the contour maps for the two homonuclear diatomics. The single covalent bond in H_2 gives rise to a relative accumulation of electrons in the bond region, between the two nuclei, at the expense of the outer, nonbonding regions of space. The triple-bond pattern for N_2 is seen to be more complex, reflecting the density accumulations in the bonding region, due to both the σ and π bonds, and the accompanying increase in the density of the lone pairs on both nitrogen atoms, due to their expected *sp*-hybridization in the atomic valence state. One also observes a decrease in the electron density in the vicinity of the nuclei and an outflow of electrons from the $2p_\pi$ AO to their overlap area, a clear sign of the orbitals involvement in the formation of the double π bond.

Both heteronuclear diatomics, HF and LiF, represent partially-ionic bonds between the two atoms exhibiting small and large differences, respectively, in their electronegativity and chemical hardness descriptors. A pattern of the density displacement in HF reflects a weakly ionic (strongly covalent) bond, while in LiF the two AIM are seen to be connected by a strongly ionic (weakly covalent) bond. Indeed, in HF one detects a relatively high degree of a "common possession" of the valence electrons by the two atoms, which significantly contribute to the shared bond-charge located between them, and a comparatively weak H→F polarization. In LiF a substantial Li→F electron transfer can be detected so that an ion-pair picture indeed provides an adequate zeroth-order description of the chemical bond in this diatomic.

Finally, in the two triatomic molecules shown in Figure 2 one identifies a strongly covalent pattern of the electron density displacements in the regions of the single N–H and C–H bonds. A typical buildup of the bond charge due to the multiple CN bonds in the two isomers HCN and HNC can be also observed. The increase in the lone-pair electron density on the terminal heavy atom, N in HCN and or C in HNC, can be also detected, thus confirming the expected *sp*-hybridization of these bonded atoms in their promoted, valence state in the molecule.

A comparison between the corresponding panels of the first two columns in the Figure 2 shows that the two displacement maps so strongly resemble one another that they are hardly distinguishable. This confirms a close relation between the local density and entropy-deficiency relaxation patterns, thus attributing to the former the complementary information-theoretic interpretation of the latter.

A strong resemblance between these two types of molecular diagrams also indicates that the local inflow of electrons increases the relative entropy, while the outflow of electrons gives rise to a diminished level of the relative-uncertainty content of the electron distribution in the molecule. The density displacement and the missing-information distribution can be thus viewed as equivalent probes of the system chemical bonds.

Similar diagnostic conclusions follow from the divergence density plots of Figure 3, where all crucial bonding and non-bonding regions of space are now identified by the *positive* values of the relative information density.

3.3. Entropy Displacement Descriptors

In Figure 4 the contour maps of the entropy-displacement density $h_\rho(\boldsymbol{r})$ are compared with the corresponding density difference diagrams for representative linear molecules of Figure 2. To better visualize details of the two functions and to facilitate a qualitative comparison between their topographies *non*-equidistant contour values have been selected. Therefore, only the profile of $h_\rho(\boldsymbol{r})$, shown in the third column of the figure, reflects the relative importance of each feature.

When interpreting these plots one should realize, that a negative (positive) value of $h_\rho(\boldsymbol{r})$ [or $h_p(\boldsymbol{r})$] signifies a decrease (increase) in the local electron *uncertainty* in the molecule, relative to the associated promolecular reference value. Again, the $\Delta\rho$ and h_ρ diagrams for H_2 are seen to qualitatively resemble one another and the corresponding map Δs shown in Figure 2. The main feature of the h_ρ diagram, an increase in the electron distribution uncertainty in the bonding region between the two nuclei, is due to the inflow of electrons to this region. This manifests the bond-covalency phenomenon, which can be attributed to the electron-sharing effect and a delocalization of the bonding electrons, now effectively moving in the field of both nuclei. One detects in all these maps a similar nodal structure and finds that the nonbonding regions exhibit a decreased uncertainty, due to a transfer of the electron density from this area to the vicinity of the two nuclei and the region between them or as a result of the orbital hybridization.

Therefore, also the molecular entropy difference function displays typical features in the reconstruction of electron distributions in a molecule, relative to those of the corresponding free atoms. Its diagrams thus provide an additional information tool for diagnosing the presence of chemical bonds through displacements in the entropy/information content of the molecular electron densities. In fact, the comparison of Figure 4 demonstrates that the entropy difference plots provide in many respects a more detailed account of the reorganization of the electronic structure, relative to the free atoms in the promolecule, compared to the corresponding density difference diagrams, particularly in the inner shell regions of heavy atoms.

In Table 1 we have listed the representative values of the molecular entropy difference of Eq. (149) together with the Shannon entropies for the molecular and promolecular electron densities. These results show that in general the molecular distribution gives rise to a lower level of the information-entropy (less electron uncertainty) compared to the promolecule. This confirms an expected higher degree of compactness of the electron distributions of bonded atoms compared to their free analogs.

Thus, the degree of uncertainty contained in the electron distribution on average decreases when the constituent free-atoms form the chemical bonds. Indeed, the dominating overall contraction of atomic electron distributions in the field of all nuclear attractors in the molecule should imply a higher degree of "order" (less uncertainty) in the molecular electron density in comparison to that present in the promolecular distribution. The largest magnitude of this relative decrease in the entropy content of the molecular electron density is observed for LiF, which exhibits the most ionic bond (largest amount of charge transfer) among all molecules included in the table.

There is no apparent correlation in Table 1 between the global entropy displacement and the bond multiplicity. For example, a triple covalent bond in N_2 generates less overall entropy loss than does a single bond in H_2. The reason for a low magnitude of the entropy displacement in N_2 is the result of a mutual cancellation of the negative and positive contributions due to valence electrons. Indeed, the orbital hybridization, AO contraction, and the charge transfer should lower the entropy of the atomic electron distribution, since they increase charge inhomogeneity in the molecule, relative to the promolecule. By the same criterion, the effective expansion of atomic densities due to the AIM promotion and the electron delocalization *via* the system chemical bonds should have the opposite effect, of relatively increasing the uncertainty content of the electron distribution in the molecule. Notice that, should one assume a similar entropy displacement of about – 0.7 bits for all triple bonds in a series of isoelectronic molecules N_2, HCN and CNH, one obtains a contribution due to a single C–H or N–H bond of about –0.7, a result close to that found for the H–H bond.

Table 1. Displacements of the molecular Shannon entropies (in bits) for molecules of Figure 2 [7,10].

Molecule	$\mathcal{H}[\rho] \equiv S[\rho] - S[\rho^0]$	$S[\rho]$	$S[\rho^0]$
H_2	– 0.84	6.61	7.45
N_2	– 0.68	8.95	9.63
HF	– 1.00	3.00	4.00
LiF	– 3.16	5.12	8.28
HCN	– 1.44	12.99	4.45
HNC	– 1.39	13.06	14.45

3.4. Illustrative Application to Propellanes

As an additional illustration we now present the combined density-difference, entropy-displacement, and information-distance analysis of the central bond in the propellane systems shown in Figure 5 [7,10]. The main purpose of this study was to examine the effect on the central C'–C' bond, between the (primed) "bridgehead" carbon atoms, of a successive increase in the size of the carbon bridges in the series of the [1.1.1], [2.1.1], [2.2.1], and [2.2.2] propellanes shown in Figure 5.

Figure 6 reports the contour maps of the molecular density difference function $\Delta\rho(\boldsymbol{r})$, the KL integrand $\Delta s(\boldsymbol{r})$, and the entropy displacement function $h_\rho(\boldsymbol{r})$, for the planes of sections displayed in Figure 5. The corresponding central-bond profiles of the density- and entropy-difference functions are shown in Figure 7. These contour maps have been generated using the DFT-LDA calculations in the extended (DZVP) basis set.

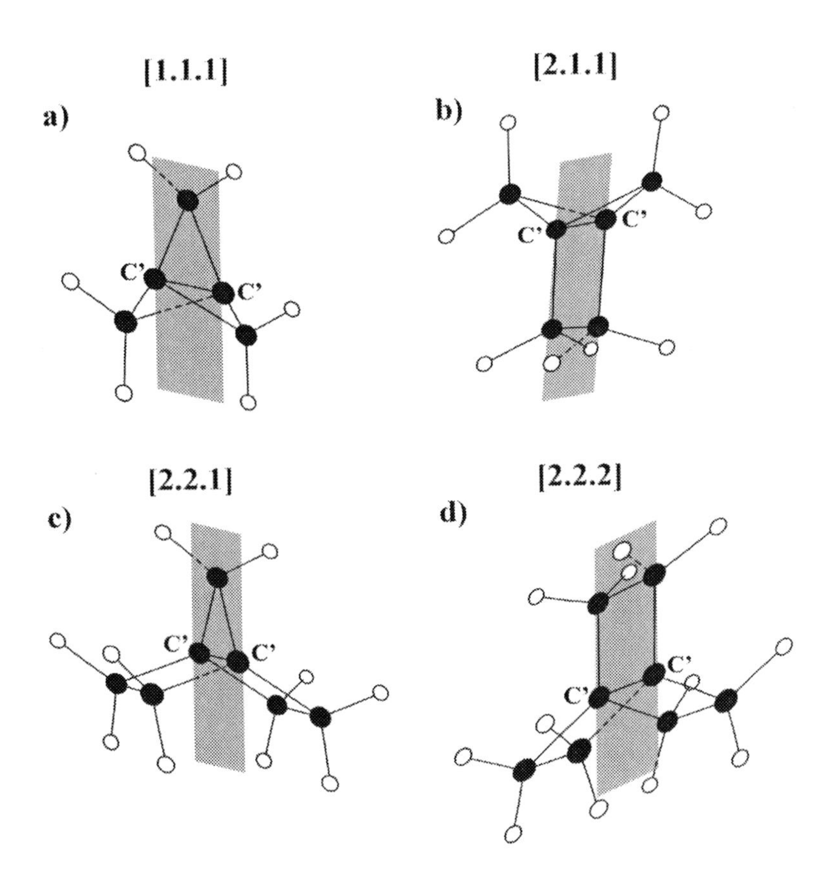

Figure 5. The propellane structures and the planes of sections containing the bridge and bridgehead (C') carbon atoms identified by black circles

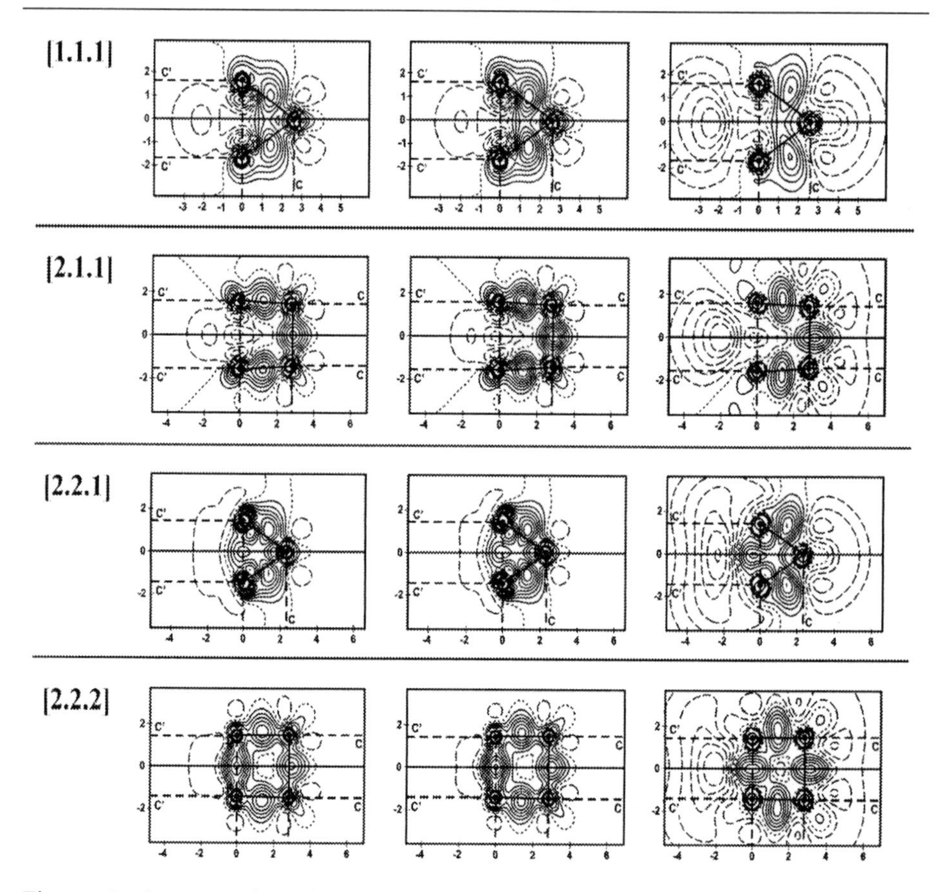

Figure 6. A comparison between the equidistant-contour maps of the density-difference function $\Delta\rho(\boldsymbol{r})$ (first column), the information-distance density $\Delta s(\boldsymbol{r})$ (second column), and the entropy-displacement density $h_\rho(\boldsymbol{r})$ (third column), for the four propellanes of Figure 5 [7,10]

The density difference plots show that in the small [1.1.1] and [2.1.1] propellanes there is on average a depletion of the electron density, relative to the promolecule, between the bridgehead carbon atoms, while the [2.2.1] and [2.2.2] systems exhibit a net density buildup in this region. A similar conclusion follows from the entropy-displacement and entropy-deficiency plots of these figures. The two entropic diagrams are again seen to be qualitatively similar to the corresponding density-difference plots. This resemblance is seen to be particularly strong between $\Delta\rho(\boldsymbol{r})$ and $\Delta s(\boldsymbol{r})$ diagrams shown in first two columns of Figure 6.

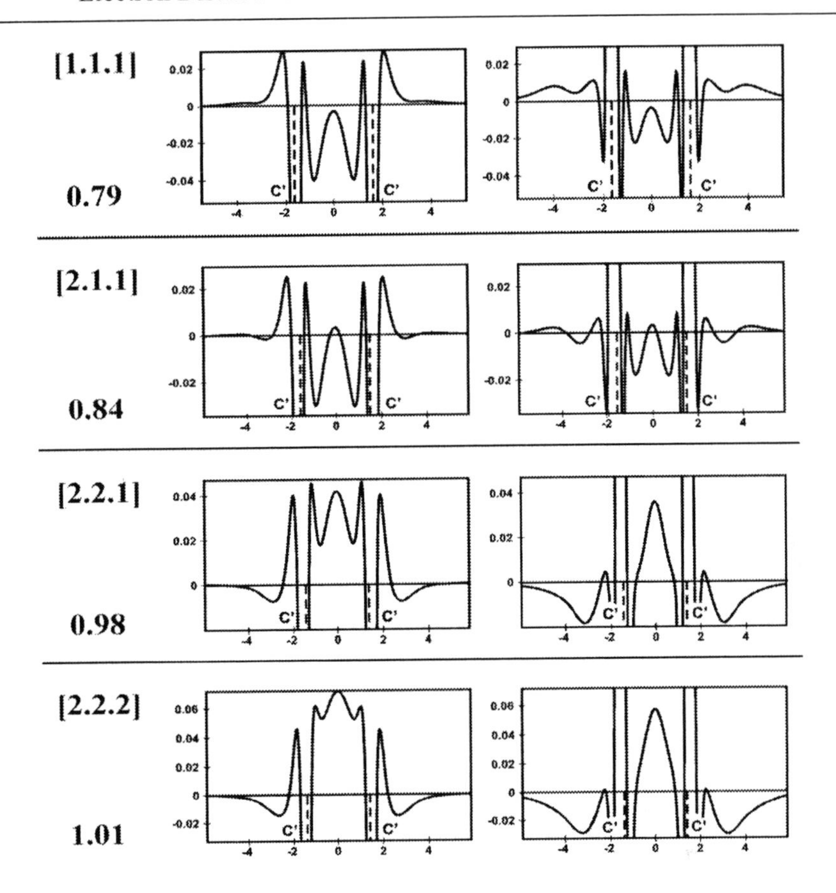

Figure 7. The bridgehead bond profiles of the density difference function (left panel) and molecular entropy displacement (right panel) for the four propellanes shown in Figure 5. For comparison the numerical values of the bond multiplicities from the difference approach [49] are also reported

A more general outlook on the bond-order concept, emerging from both the quadratic-difference approache in MO theory [46-52] and from OCT [8,20-22,24], identifies the chemical bond as a measure of a statistical "*dependence*" (non-additivity) between orbitals on different atomic centers. On one hand, this dependence between basis functions of different atoms is partly realized *directly* (through space), by a constructive interference of orbitals (probability amplitudes) on two atoms, which increases the electron density between them. On the other hand, it has an *indirect* origin, through the depemdence on orbitals of the remaining AIM. The latter is due to the orthonormality relations of the occupied MO, which determine the framework of chemical bonds.

The numerical bond-orders reported in Figure 7 originate from the *two*-electron difference approach [46-52], an extension of the original (covalent) bond multiplicity index of Wiberg [39] and its numerous (covalent) generalizations, e.g., [53,54]. Together with the corresponding density profiles shown in this figure they reveal a changing nature of the central bond in the four propellanes included in this analysis. The central "bonds" in the smallest systems, lacking the accumulation of the electron density or the entropy/entropy-deficiency density between the bridgehead atoms, are seen to be mostly of the *indirect* character, being realized "through-bridges" rather than directly "through-space".

Clearly, the most important atomic intermediates for this indirect bond mechanism in propellanes are the *bridge* carbons, which strongly overlap with the *bridgehead* carbons. A gradual emergence of the direct, "through-space" component of the central bond, due to accumulation of the electron density and the entropy (entropy-deficiency) density between the bridgehead carbons, is observed when the bridges are enlarged in the two largest propellanes. Using the two-electron difference approach one roughly estimates [49] a full single bond in the [2.2.1] and [2.2.2] propellanes and approximately 0.8 bond-order in the [1.1.1] propellane. Hence, using the latter estimate as a measure of the "through-bridges" component in the largest propellane, one predicts about 0.2 bond-order measure for the "through-space" component of the central bond in the largest [2.2.2] propellane.

The missing through-space component in the smallest [1.1.1] system is due to nearly tetrahedral (sp^3) hybridization on the bridge-head carbons, with the three hybrids on each of these atoms being used to form the chemical bonds with the bridge carbons and the fourth hybrid being directed away from the other bridgehead position. In the largest [2.2.2] propellane these central carbons acquire a nearly trigonal (sp^2) hybridization, to form bonds with the bridge neighbors, with the third $2p$ orbital now being directed along the bridgehead axis, thus being capable of forming a strong through-space component of the overall bond multiplicity.

Chapter 4

BONDED ATOMS FROM INFORMATION THEORY

4.1. CHEMICAL CONCEPTS

In chemistry an understanding of the electronic structure of molecules and their preferences in reactions comes from transforming the computed (or experimental) electron distributions into statements in terms of chemical concepts, such as bonded atoms, the building blocks of molecules, their collections defining larger fragments of the whole molecular system, e.g., the functional groups, and chemical bonds, which represent the molecular "connectivities" between atoms, and other molecular subsystems, e.g., σ and π electrons in benzene. Indeed, an understanding of molecules as combinations of atoms is fundamental to chemistry. It is not surprising, then, that the concept of AIM has been much discussed in scientific literature, e.g., [7,15,55].

In fact, chemistry deals mainly with rather small changes of bonded atoms and larger molecular fragments, with reasonably well understood and *transferable* molecular invariants, such as AIM, functional groups, molecular subsystems, e.g., reactants and products of an elementary chemical reaction, etc., which tend to maintain their identity in different molecular environments. Most molecular systems may be thought of as consisting of only slightly perturbed atoms (or atomic ions), deformed by the presence of the molecular remainder and exhibiting somewhat modified net electric charges. These displacements in the atomic electronic structure, relative to the corresponding free-atom states, are due to the coupled processes of the *intra*-atomic *Polarization* (P), responsible for the *promotion* of bonded atoms to their effective *valence-states* in the molecule, and the *inter*-atomic *Charge Transfer* (CT), which accompany the formation of the system chemical bonds.

An important part of the chemical science is based on such intuitive notions, which ultimately escape the rigorous definition in molecular quantum mechanics, thus representing the Kantian *noumenons* of chemistry [15]. The natural question to be addresses is: can IT help in making these concepts more precisely defined in terms of the molecular and/or promolecular electron probabilities, which together reflect the information redistribution during the bond formation process?

The bonded fragments of a given molecular system represent the mutually-*open* subsystems capable of exchanging electrons with their respective molecular reminders. One would hope to find that a given AIM, like its free (non-bonded) analog, would possess a single cusp at the nucleus in its electron density, linked to the effective atomic number of the nucleus. The bonded atoms are pieces of the molecule so that their electron densities $\{\rho_i\}$ must sum up to the molecular electron density $\rho = \sum_i \rho_i$. However, since each bonded atom preserves to a remarkably high degree the free-atom identity, the AIM distributions should be also closely related to electron densities $\{\rho_i^0\}$ of the separated atoms, which define the atomic promolecule consisting of such *free*-atom densities shifted to the actual AIM locations in the molecule. The resulting electron density, $\rho^0 = \sum_i \rho_i^0$, of this collection of the "frozen" atomic electron distributions, which defines the initial stage in the bond formation process, thus determines a natural reference for extracting changes due to the chemical bonds. Indeed, the familiar density difference function of Chapter 3, $\Delta\rho = \rho - \rho^0$, has been widely used to probe the chemical bonds in molecules.

The electronic structure of molecular systems is characterized by their *one-*, *two-* and *many*-electron probability distributions in the continuous ("fine-grained") description. To obtain its *chemical* interpretation, e.g., in terms of AIM, functional groups, reactants or other type of chemically significant subsystems, e.g., the σ and π electrons in aromatic compounds, these overall distributions have to be "discretized" in terms of the relevant pieces of the overall density attributed to the constituent parts of the molecular system under consideration, e.g., the bonded atoms or molecular fragments (clusters of AIM). The densities of molecular subsystems constitute their fine-grained description. By an appropriate integration of the electron/probability densities one then obtains the corresponding *condensed* descriptors of the electronic structure of molecules and their fragments, providing the associated discrete (*coarse-grained*) indices of the underlying continuous electron distributions. Additional resolution levels are provided by the

AO or MO representations resulting from the quantum-mechanical calculations of the molecular electronic structure.

It should be emphasized, however, that an exhaustive partitioning of the given molecular electron density between constituent (bonded) atoms, which determines the AIM effective net charges (oxidation states) in a given molecular environment, is not unique, since it depends on the adopted criterion for such a division [7,15]. The imposing variety of published theoretical methods for partitioning the molecular density into *"best"* AIM contributions testifies to the importance of this theme in chemistry. Different methods are based on different principles, some to a degree arbitrary or heuristic, which can produce conflicting trends in the associated AIM charges. Methods differ in the theoretical techniques used, e.g., topological analysis of the density, wave-function description, or the density-functional approach. They also differ in the physical/heuristic principles invoked, e.g., electronegativity equalization, zero flux, the minimum promotion-energy rules, and the minimum entropy-deficiency (information-distance, missing-information) criteria in the IT approach.

The historically first scheme of the Mulliken/Löwdin *population analysis* have used the *function* space partitioning, in which one distributes electrons between AO which form the basis set for expanding MO of the Hartree-Fock (SCF LCAO MO) theory and its *Configuration Interaction* (CI) extension, non-orthogonal in the Mulliken approach and the symmetrically orthogonalized in the Löwdin variant. Another popular approach of Bader [55], with a solid topological and quantum-mechanical basis, uses the *physical* space partitioning, i.e., a division of space into the exclusive atomic basins, with the boundaries determined by the zero-flux surfaces, on which the flow of electrons between subsystems vanishes. In the latter approach the spherical, spatially non-confined bonded atoms of the population analysis are replaced by the topological, non-spherical pieces of the molecular density, obtained as cuts along the zero-flux surfaces. As a result the topological AIM represent the spatially-confined (non-overlapping) and strongly non-symmetrical atoms.

4.2. Stockholder Atoms-in-Molecules

Yet another, *stockholder* division scheme of Hirshfeld [23], which has been widely exploited in crystallography, uses the "common-sense" *local* partitioning principle, which parallels the familiar stock market rule: in forming a molecule each bonded atom locally partakes of the molecular density ("profit") gain or loss in the molecule in proportion to its share in the promolecular density ("investment"). Thus, by construction, the (overlapping) stockholder AIM are infinitely extending (not confined spatially) and known to be only slightly polarized relative to the free-atom reference. Both the topological and stockholder atoms are derived from the molecular electron density, and as such they preserve a "memory" of the original overall distribution of electrons in the molecule. This *one*-electron stockholder division scheme has recently been shown to have a strong basis in IT [7,11-17]. However, the *two*-electron generalization of the stockholder principle [13] generates in molecules involving light atoms, e.g., hydrogen and lithium, slightly different effective one-electron distributions of the associated bonded hydrogens, which emphasize more strongly the bonding (overlap) regions between the bond partners [17] compared to the corresponding *one*-electron stockholder atoms of Hirshfeld.

It has been shown by Hirshfeld [23] that the electron density $\rho(\boldsymbol{r})$ of the molecular system M = $(A^{H} \mid B^{H} \mid \ldots)$, consisting of the mutually-*open* atoms $X^{H} = (A^{H}, B^{H}, \ldots)$, as marked by the perpendicular *broken* lines separating the AIM symbols in M, can be exhaustively partitioned, into the "stockholder" AIM densities $\{\rho_X^{H}(\boldsymbol{r})\} \equiv \boldsymbol{\rho}^{H}(\boldsymbol{r})$:

$$\begin{aligned}
\rho_X^{H}(\boldsymbol{r}) &= \rho_X^{0}(\boldsymbol{r})\,[\rho(\boldsymbol{r})/\rho^{0}(\boldsymbol{r})] \equiv \rho_X^{0}(\boldsymbol{r})\,w(\boldsymbol{r}) \\
&= \rho(\boldsymbol{r})\,[\rho_X^{0}(\boldsymbol{r})/\rho^{0}(\boldsymbol{r})] \equiv \rho(\boldsymbol{r})\,d_X^{H}(\boldsymbol{r}), \qquad X = A, B, \ldots \\
&\qquad \Sigma_X\, d_X^{H}(\boldsymbol{r}) = 1, \qquad \rho(\boldsymbol{r}) = \Sigma_X\, \rho_X^{H}(\boldsymbol{r}). \qquad (154)
\end{aligned}$$

Here $\boldsymbol{\rho}^{0}(\boldsymbol{r}) = \{\rho_X^{0}(\boldsymbol{r})\}$ groups the densities of the free constituent atoms, giving rise to the reference electron density $\rho^{0}(\boldsymbol{r}) = \Sigma_X \rho_X^{0}(\boldsymbol{r})$ of the (isoelectronic) promolecule $M^{0} = (A^{0} | B^{0} | \ldots)$, consisting of the non-bonded (mutually-*closed*) atoms $X^{0} = (A^{0}, B^{0}, \ldots)$, as marked by the perpendicular *solid* lines separating the free atom symbols in M^{0}:

$$\begin{aligned}
\int \rho^{0}(\boldsymbol{r})\, d\boldsymbol{r} &= \Sigma_X \int \rho_X^{0}(\boldsymbol{r})\, d\boldsymbol{r} = \Sigma_X N_X^{0} = N^{0} \\
&= \int \rho(\boldsymbol{r})\, d\boldsymbol{r} = \Sigma_X \int \rho_X^{H}(\boldsymbol{r})\, d\boldsymbol{r} = \Sigma_X N_X^{H} = N. \qquad (155)
\end{aligned}$$

The free-atom densities $\boldsymbol{\rho}^{0}$ in M^{0} are shifted to the respective atomic positions in the molecule and the vectors $\boldsymbol{N}^{H} = \{N_X^{H}\}$ and $\boldsymbol{N}^{0} = \{N_X^{0}\}$ group the atomic average numbers of electrons of the bonded and free atoms, respectively. As we have already observed in Section 3.1, the same promolecular reference is used to determine the density difference function $\Delta\rho(\boldsymbol{r}) = \rho(\boldsymbol{r}) - \rho^{0}(\boldsymbol{r})$.

A reference to Eq. (154) shows that the Hirshfeld AIM densities satisfy the local principle of the *one*-electron stockholder division, which can be stated as the following equality between the local molecular and promolecular *conditional* probabilities:

$$d_X^H(\boldsymbol{r}) = \rho_X^H(\boldsymbol{r})/\rho(\boldsymbol{r}) \equiv P^H(X|\boldsymbol{r})$$
$$= d_X^0(\boldsymbol{r}) = \rho_X^0(\boldsymbol{r})/\rho^0(\boldsymbol{r}) \equiv P^0(X|\boldsymbol{r}), \quad \sum_X P^H(X|\boldsymbol{r}) = \sum_X P^0(X|\boldsymbol{r}) = 1. \quad (156)$$

As we have also remarked above, this relation has been interpreted by Hirshfeld using the stock market analogy: each atom participates locally in the molecular "profit" $\rho(\boldsymbol{r})$ in proportion to its "share" $d_X^0(\boldsymbol{r}) = P^0(X|\boldsymbol{r})$ in the promolecular "investment" $\rho^0(\boldsymbol{r})$. In the remaining part of this section we shall demonstrate that this common-sense division rule has a solid basis in IT [7,11-16].

By extracting the overall number of electrons $N = N^0$ from the molecular and subsystem densities one introduces the associated (molecularly-normalized) probability distributions:

$$\rho(\boldsymbol{r}) = Np(\boldsymbol{r}) = N\sum_X p_X^H(\boldsymbol{r}) \quad \text{and} \quad \boldsymbol{\rho}^H(\boldsymbol{r}) = N\boldsymbol{p}^H(\boldsymbol{r}) = N\{p_X^H(\boldsymbol{r})\}. \quad (157)$$

Here $p(\boldsymbol{r})$ and $\boldsymbol{p}^H(\boldsymbol{r})$ stand for the *shape*-factors of the system as a whole and of its Hirshfeld atoms, respectively,

$$\sum_X \int p_X^H(\boldsymbol{r})\,d\boldsymbol{r} = \sum_X (N_X^H/N) \equiv \sum_X P_X^H = 1, \quad (158)$$

while $\boldsymbol{P}^H = \{P_X^H\}$ groups the condensed probabilities of finding an electron of M on the specified stockholder AIM.

This molecular normalization reflects the important fact that bonded atoms are constituent parts of the molecule, so that the full normalization condition has to involve the summation/integration over the complete set of *one*-electron events, consisting of all possible "values" of the discrete argument X (atomic label) and all spatial locations of an electron, identified by continuous coordinates $\boldsymbol{r}$ in the subsystem probability distributions:

$$\boldsymbol{p}^H(\boldsymbol{r}) = \{p_X^H(\boldsymbol{r}) \equiv P^H(X \wedge \boldsymbol{r}) = p(\boldsymbol{r})P^H(X|\boldsymbol{r})\}. \quad (159)$$

The same, global normalization has to be adopted for the *free*-atom pieces of the *one*-electron probability distribution in the isoelectronic promolecule and for its *free*-atom components, respectively,

$$p^0(\boldsymbol{r}) = \rho^0(\boldsymbol{r})/N = \sum_X p_X{}^0(\boldsymbol{r}), \qquad \boldsymbol{p}^0(\boldsymbol{r}) = \boldsymbol{\rho}^0(\boldsymbol{r})/N = \{p_X{}^0(\boldsymbol{r})\};$$

$$\sum_X \int p_X{}^0(\boldsymbol{r})\, d\boldsymbol{r} = \sum_X (N_X{}^0/N) \equiv \sum_X P_X{}^0 = 1, \tag{160}$$

where $\boldsymbol{P}^0 = \{P_X{}^0\}$ collects the condensed probabilities of observing an electron of M^0 on the specified free atom. Again the full normalization of the shape (probability) factors $\boldsymbol{p}^0(\boldsymbol{r}) = \{p_X{}^0(\boldsymbol{r}) \equiv P^0(X \wedge \boldsymbol{r})\}$ of the non-bonded atoms in the promolecular system involves summation over the discrete atomic "variable" X and integration over all positions $\boldsymbol{r}$ of an electron, the latter representing a continuous-event label of the probability distributions of atomic fragments in M^0.

It also follows from Eq. (154) that in the *one*-electron stockholder-division scheme each free subsystem density (or its shape factor) is locally modified in accordance with the *molecular* (subsystem-independent) *enhancement factor* $w(\boldsymbol{r})$:

$$w_X{}^H(\boldsymbol{r}) \equiv \rho_X{}^H(\boldsymbol{r})/\rho_X{}^0(\boldsymbol{r}) = p_X{}^H(\boldsymbol{r})/p_X{}^0(\boldsymbol{r})$$

$$= \rho(\boldsymbol{r})/\rho^0(\boldsymbol{r}) = p(\boldsymbol{r})/p^0(\boldsymbol{r}) \equiv w(\boldsymbol{r}). \tag{161}$$

Therefore, this procedure is devoid of any subsystem bias and as such appears to be fully objective.

Representative plots of the overlapping electron densities of the bonded stockholder hydrogens in H_2 are shown in Figure 8. They are seen to be distributed all over the physical space, decaying exponentially at large distances from the molecule and exhibiting in the bond density profile a single cusp at the atomic nucleus. They also display the expected polarization towards the bonding partner. These subsystem densities are highly transferable [23] and their overlap in the molecule accords with the classical interpretation of the origin of the chemical bond. One also observes a higher AIM density at the atomic nucleus, in comparison to the free-hydrogen density, i.e., a contraction of the AIM distribution at the expense of the nonbonding part of the free-atom density. This is due to a presence of the other atom causing an effective lowering of the molecular external potential relative to the external potential of the separated atom.

Let us briefly examine the asymptotic properties of the stockholder atomic densities. For simplicity we consider a diatomic system $M = (A^H|B^H)$, $\rho = \rho_A{}^H + \rho_B{}^H$, consisting of two Hirshfeld atoms A^H and B^H, the free analogs of which, A^0 and B^0, exhibit the relative electron *acceptor* (acidic) and *donor* (basic) properties, respectively. This assumption implies $I_A{}^0 > I_B{}^0$, where $I_X{}^0$ denotes the *ionization potential* of X^0. Rewriting Eq. (154) in terms of the local density ratio $x = \rho_B{}^0/\rho_A{}^0$ gives:

$$\rho_A^H = (1 + x)^{-1}\rho \qquad \text{and} \qquad \rho_B^H = (1 + x^{-1})^{-1}\rho. \tag{162}$$

Hence, for $r_X = |\boldsymbol{r} - \boldsymbol{R}_X| \to \infty$, when the distances from both nuclei become large compared to the inter-atomic separation R_{AB}, $x \to \infty$ since the asymptotic behavior of the free subsystems of the promolecule are determined by the their ionization potentials, i.e., the negative energies of the highest occupied KS orbitals:

$$\rho_X^0 \to \exp[-2(2I_X^0)\, r_X]\ (r_X \to \infty), \qquad X = A, B. \tag{163}$$

Therefore, for $r_X \to \infty$, $\rho_A^H \to 0$ and $\rho_B^H \to \rho$, so that the density of the softer (donor) atom B has a dominant contribution to the molecular density at distances from the molecule large compared to R_{AB}.

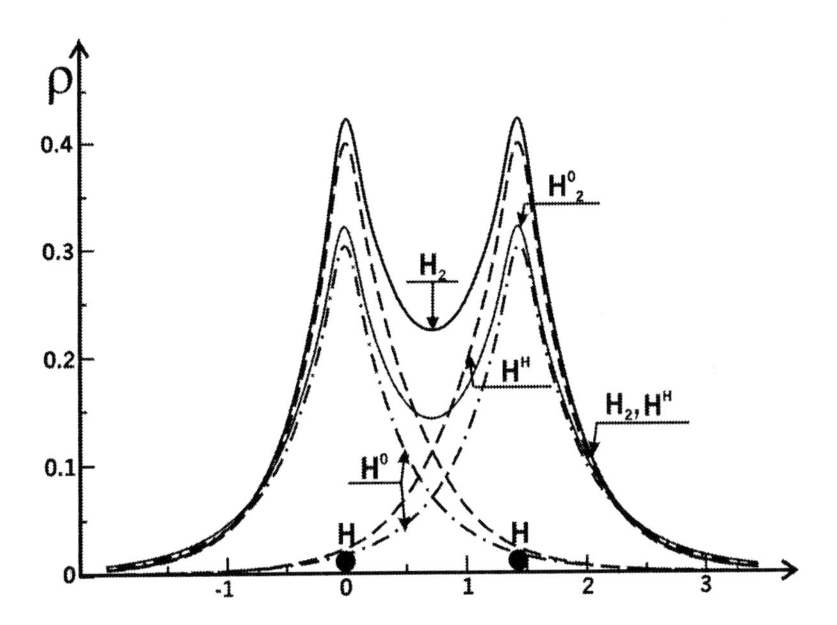

Figure 8. The Hirshfeld electron densities of bonded hydrogen atoms (H^H) obtained from the molecular density (H_2). The free-hydrogen densities (H^0) and the resulting electron density of the promolecule (H_2^0) are also shown for comparison. The density and inter-nuclear distance are in a.u. The zero cusps at nuclear positions are the artifacts of the Gaussian basis set used in DFT calculations

Table 2. Representative net charges $\{q_X^H = N_X^H - Z_X\}$ (a.u.) of the stockholder AIM, where Z_X denotes the charge of nucleus X, the AIM entropy deficiencies $\{\Delta S_X[p_X^H|p_X^0]\}$, and the global entropy deficiency $\Delta S[p|p^0]$ (in bits), for the linear molecules of Figure 2 [9]

Molecule	X	q_X^H	$\Delta S_X[p_X^H|p_X^0]$	$\Delta S[p|p^0]$
H_2	H	0.00	0.056	0.056
N_2	N	0.00	0.006	0.006
HF	H	0.24	0.144	0.020
	F	− 0.24	0.005	
LiH	Li	0.35	0.157	0.136
	H	− 0.35	0.012	
LiF	Li	0.58	0.244	0.063
	F	− 0.58	0.007	
LiCl	Li	0.53	0.212	0.033
	Cl	− 0.53	0.003	
HCN	H	0.14	0.104	0.017
	C	0.03	0.015	
	N	− 0.17	0.005	
HNC	H	0.20	0.110	0.018
	N	− 0.10	0.008	
	C	− 0.10	0.011	
HNCS	H	0.19	0.114	0.008
	N	− 0.13	0.007	
	C	0.05	0.008	
	S	− 0.11	0.002	
HSCN	H	0.22	0.088	0.008
	S	− 0.07	0.002	
	C	0.04	0.008	
	N	− 0.19	0.004	

In Table 2 we have listed the net charges and entropy-deficiencies of the Hirshfeld AIM for illustrative linear molecules included in Figure 2 [7,9]. It follows from these numerical results that the Hirshfeld charges represent the chemical intuition quite well. For example, in the series of heteronuclear diatomics of increasing bond ionicity due to a growing electronegativity difference between constituent atoms, HF, LiH, LiCl, and LiF, the amount of CT monotonically increases, as intuitively expected.

The atomic missing information,

$$\Delta S_X[p_X^H|p_X^0] = \int p_X^H(\boldsymbol{r}) \log[p_X^H(\boldsymbol{r})/p_X^0(\boldsymbol{r})]\, d\boldsymbol{r}\,, \tag{164}$$

reflects the information distance between the atomic shape-factors $p_X^H(\boldsymbol{r})$ and $p_X^0(\boldsymbol{r})$. The reported values of these quantities are quite small, thus numerically confirming that the bonded atoms do indeed strongly resemble their free-atom analogs. The same general conclusion follows from examining the reported global entropy deficiencies:

$$\Delta S[p\,|\,p^0] = \Sigma_X\, \Delta S_X[p_X^H|p_X^0] = \int p(\boldsymbol{r}) \log[p(\boldsymbol{r})/p^0(\boldsymbol{r})]\, d\boldsymbol{r} \equiv \int p(\boldsymbol{r})\, I[w(\boldsymbol{r})]\, d\boldsymbol{r}. \tag{165}$$

This strong similarity between the molecular and promolecular electron distributions is also seen in Figure 8: the appreciable changes of the free-atom densities in the molecule are only observed around the nuclei (a contraction of the *free*-atom density) and in the bond region between the two nuclei (a polarization of the free atoms towards the bonding partner). As expected, in heavier atoms only a slight distortion of the valence (external) electrons is observed in the stockholder (bonded) atoms, with the inner shell structure left practically intact.

4.3. Information-Theoretic Justification

The optimum *local* partition of the molecular density can be best formulated in terms of the unknown *conditional* probabilities $\boldsymbol{d}(\boldsymbol{r}) = \{d_X(\boldsymbol{r}) \equiv P(X|\boldsymbol{r})\}$, which uniquely determine the AIM pieces of $p(\boldsymbol{r})$, $\{p_X(\boldsymbol{r}) = d_X(\boldsymbol{r})p(\boldsymbol{r})\}$ [7,13,14]. The relevant (local) KL function of the unknown share factors $\boldsymbol{d}(\boldsymbol{r})$, which measures the local information distance relative to the promolecular reference values $\boldsymbol{d}^0(\boldsymbol{r}) = \{P^0(X|\boldsymbol{r})\}$, is then given (in nats) by the sum of AIM contributions:

$$\Delta S[\boldsymbol{d}(\boldsymbol{r})|\boldsymbol{d}^0(\boldsymbol{r})] = \Sigma_X\, P(X|\boldsymbol{r}) \ln[P(X|\boldsymbol{r})/P^0(X|\boldsymbol{r})]. \tag{166}$$

The best (unbiased) share factors of subsystems, giving rise to the maximum similarity of the bonded fragment densities to their non-bonded (free) analogs, must minimize this missing information function subject to the normalization constraint of the local conditional probabilities, $\sum_X P(X|\boldsymbol{r}) = 1$,

$$\delta\{\Delta S[\boldsymbol{d}(\boldsymbol{r})|\boldsymbol{d}^0(\boldsymbol{r})] - \lambda(\boldsymbol{r}) \sum_X P(X|\boldsymbol{r})\} = 0, \tag{167}$$

where $\lambda(\boldsymbol{r})$ stands for the Lagrange multiplier associated with this auxiliary condition at the given location $\boldsymbol{r}$. The resulting Euler equation for the optimum local conditional probability $P^{opt.}(X|\boldsymbol{r})$,

$$\ln[P^{opt.}(X|\boldsymbol{r})/P^0(X|\boldsymbol{r})] + [1 - \lambda(\boldsymbol{r})] \equiv \ln\{P^{opt.}(X|\boldsymbol{r})/[CP^0(X|\boldsymbol{r})]\} = 0, \tag{168}$$

or $P^{opt.}(X|\boldsymbol{r}) = CP^0(X|\boldsymbol{r})$, when combined with the normalization constraint, $\sum_X P^{opt.}(X|\boldsymbol{r}) = C\sum_X P^0(X|\boldsymbol{r}) = C = 1$, is then seen to give the Hirshfeld solution of Eqs. (154) and (161): $\boldsymbol{d}^{opt.}(\boldsymbol{r}) = \boldsymbol{d}^0(\boldsymbol{r}) = \boldsymbol{d}^H(\boldsymbol{r})$. Therefore, the Hirshfeld choice of the local share factors minimizes the local information distance of Eq. (166) to the lowest value possible: $\Delta S[\boldsymbol{d}^H(\boldsymbol{r})|\boldsymbol{d}^0(\boldsymbol{r})] = 0$.

The same answer follows from the alternative, *global* information principles formulated in terms of either the electron densities or their shape factors [11], in which one seeks the optimum overall atomic (or fragment) distributions exhibiting the strongest resemblance to the corresponding non-bonded, reference distribution of the associated promolecular system. Let us define the KL information-distance functional (see Section 1.2) between the trial *one*-electron densities of atomic fragments $\{\rho_X\} \equiv \boldsymbol{\rho}$ (or the associated probability distributions $\{p_X\} \equiv \boldsymbol{p} = \boldsymbol{\rho}/N$) of the bonded atoms and the corresponding reference densities $\{\rho_X^0\} \equiv \boldsymbol{\rho}^0$ (or $\{p_X^0\} \equiv \boldsymbol{p}^0 = \boldsymbol{\rho}^0/N$) of the free-atoms:

$$\begin{aligned}\Delta S[\boldsymbol{\rho}|\boldsymbol{\rho}^0] &= \sum_X \int \rho_X(\boldsymbol{r}) \ln[\rho_X(\boldsymbol{r})/\rho_X^0(\boldsymbol{r})]\, d\boldsymbol{r} \equiv \sum_X \int \rho_X(\boldsymbol{r})\, I_X[w_X(\boldsymbol{r})]\, d\boldsymbol{r} \\ &\equiv \sum_X \Delta S_X[\rho_X|\rho_X^0] \\ &= N\sum_X \int p_X(\boldsymbol{r}) \ln[p_X(\boldsymbol{r})/p_X^0(\boldsymbol{r})]\, d\boldsymbol{r} \equiv N\sum_X \int p_X(\boldsymbol{r})\, I_X[w_X(\boldsymbol{r})]\, d\boldsymbol{r} \\ &\equiv N\Delta S[\boldsymbol{p}|\boldsymbol{p}^0] \equiv N\sum_X \Delta S_X[p_X|p_X^0].\end{aligned} \tag{169}$$

Here, $I_X[w_X(\boldsymbol{r})] = \ln[p_X(\boldsymbol{r})/p_X^0(\boldsymbol{r})] = \ln[\rho_X(\boldsymbol{r})/\rho_X^0(\boldsymbol{r})]$, stands for (variational) surprisal of atom X, for the current value of the local enhancement factor relative to the *free*-subsystem reference: $w_X(\boldsymbol{r}) = \rho_X(\boldsymbol{r})/\rho_X^0(\boldsymbol{r}) = p_X(\boldsymbol{r})/p_X^0(\boldsymbol{r})$.

We have also indicated in the preceding equation that the directed divergence of $\boldsymbol{\rho}$ relative to $\boldsymbol{\rho}^0$, $\Delta S[\boldsymbol{\rho}|\boldsymbol{\rho}^0]$ is just N times the entropy deficiency $\Delta S[\boldsymbol{p}|\boldsymbol{p}^0]$. The same relation holds between the subsystem missing-information quantities: $\Delta S_X[\rho_X|\rho_X^0] = N\ \Delta S_X[p_X|p_X^0]$. Therefore, for the isoelectronic molecular and promolecular systems, the problem of normalization of the compared electronic densities does not influence the corresponding constrained variational principle for determining the optimum densities of atomic pieces:

$$\delta\{\Delta S[\boldsymbol{\rho}|\boldsymbol{\rho}^0] - \int\lambda(\boldsymbol{r})\Sigma_X\,\rho_X(\boldsymbol{r})d\boldsymbol{r}\} = N\delta\{\Delta S[\boldsymbol{p}|\boldsymbol{p}^0] - \int\lambda'(\boldsymbol{r})\Sigma_X\,p_X(\boldsymbol{r})d\boldsymbol{r}\} = 0, \quad (170)$$

where the local Lagrange multipliers $\lambda(\boldsymbol{r})$ and $\lambda'(\boldsymbol{r})$ enforce the corresponding conditions of the exhaustive-division at point $\boldsymbol{r}$: $\Sigma_X\rho_X(\boldsymbol{r}) = \rho(\boldsymbol{r})$ and $\Sigma_X\,p_X(\boldsymbol{r}) = p(\boldsymbol{r})$, respectively. The above variational problem in terms of electron densities is indeed seen to be equivalent to the associated principle in terms of the probability distributions (shape-factors):

$$\delta\{\Delta S[\boldsymbol{p}|\boldsymbol{p}^0] - \int\lambda'(\boldsymbol{r})\,\Sigma_X\,p_X(\boldsymbol{r})\,d\boldsymbol{r}\} = 0, \quad (171)$$

with $\lambda'(\boldsymbol{r})$ now multiplying the local exhaustive-division constraint: $\Sigma_X\,p_X(\boldsymbol{r}) = p(\boldsymbol{r})$.

It can be easily verified by a straightforward functional differentiation that both these variational principles give the same answer of the local stockholder division: $p_X^{opt.}(\boldsymbol{r}) = p_X^H(\boldsymbol{r}) = \rho_X^H(\boldsymbol{r})/N$. In these global minimum entropy-deficiency principle the missing-information term provides an entropy "penalty" for the AIM densities deviating from the corresponding free-atom densities.

This Hirshfeld solution of the IT density-partitioning problem is thus independent of the adopted entropy/information measure. The symmetrized divergence of Kullback,

$$\Delta S[\boldsymbol{\rho}^0, \boldsymbol{\rho}] = \Delta S[\boldsymbol{\rho}|\boldsymbol{\rho}^0] + \Delta S[\boldsymbol{\rho}^0|\boldsymbol{\rho}] = \Sigma_X \int\Delta\rho_X(\boldsymbol{r})\ln[\rho_X(\boldsymbol{r})/\rho_X^0(\boldsymbol{r})]\,d\boldsymbol{r}$$

$$= N\Delta S[\boldsymbol{p}^0, \boldsymbol{p}], \quad (172)$$

where $\Delta\rho_X(\boldsymbol{r}) = \rho_X(\boldsymbol{r}) - \rho_X^0(\boldsymbol{r}) = N\Delta p_X(\boldsymbol{r})$, gives rise to the same optimum division of the molecular electron distribution into AIM fragments. The Fisher intrinsic accuracy generates the differential Euler equation for the optimum subsystem distributions. One of its particular solutions is also the Hirshfeld prescription for the bonded atom pieces of the molecular electron density.

The minimum entropy-deficiency principle can be similarly applied to a division of the molecular *joint* distribution of k electrons into the corresponding pieces describing the AIM k-clusters. This approach gives rise to the associated *many*-electron stockholder principle [7,13,16,17]. By an appropriate partial summation and integration these cluster distriubutions can be used to determine the associated effective *one*-particle densities $\{\rho_X^{eff.}(\boldsymbol{r};\ k)\}$ of these k-electron "stockholder" (k-S) atoms, which in general slightly differ from their *one*-electron (1-S) analogs of Hirshfeld. The largest differences (see Figure 9) between densities of the 2-S and 1-S bonded atoms have been observed for the lightest hydrogen and lithium atoms [7,17]. As seen in the figure the 2-S H and Li atoms exhibit a more bonding (overlapping) character, compared to their 1-S analogs.

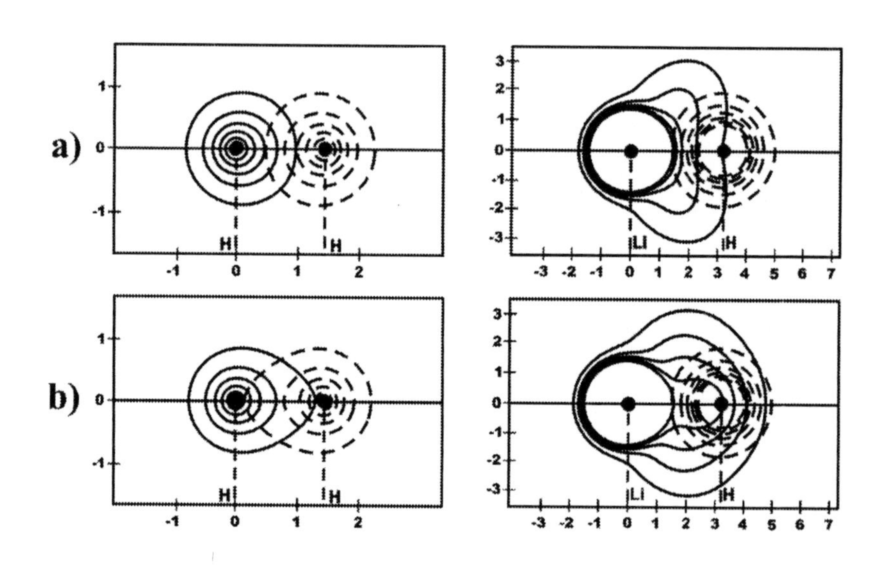

Figure 9. A comparison of the contour diagrams of the $\rho_X^{H}(\boldsymbol{r})$ (1-S, Panel a) and $\rho_X^{eff}(\boldsymbol{r})$ (2-S, Panel b) distributions of bonded atoms in H_2 (left column) and LiH (right column) [7,17]

4.4. Representative Information Densities

The same entropic descriptors, which have been used in Sections 3.1 and 3.2 to probe the molecular electron densities/probabilities, can be applied to diagnose changes the bonded atoms undergo in molecules, relative to the free atoms [7,10]. The entropy deficiency of the stockholder AIM,

$$\Delta S[\rho_X^H|\rho_X^0] = \int \rho_X^H(\boldsymbol{r}) \log[\rho_X^H(\boldsymbol{r})/\rho_X^0(\boldsymbol{r})]\, d\boldsymbol{r} \equiv \int \rho_X^H(\boldsymbol{r})\, I_X^H(\boldsymbol{r})$$

$$= \int P^H(X|\boldsymbol{r})\rho(\boldsymbol{r})\, I(\boldsymbol{r})\, d\boldsymbol{r} \equiv \int P^H(X|\boldsymbol{r})\, \Delta s(\boldsymbol{r})\, d\boldsymbol{r} \equiv \int \Delta s_X^H(\boldsymbol{r})\, d\boldsymbol{r},$$

$$\sum_X P^H(X|\boldsymbol{r}) = 1, \tag{173}$$

defines the density of the atomic cross-entropy (shown in Figs. 10 and 11), $\Delta s_X^H(\boldsymbol{r}) = P^H(X|\boldsymbol{r})\Delta s(\boldsymbol{r})$, where the conditional probability $P^H(X|\boldsymbol{r})$ is defined by the stockholder share-factor $d_X^H(\boldsymbol{r})$ of Eq. (156), and the atomic surprisals $\{I_X^H(\boldsymbol{r})\}$ are all equalized at the global (molecular) value: $I(\boldsymbol{r}) = I(w(\boldsymbol{r}))$ [see Eq. (161)].

The global values of the entropy displacement of the Hirshfeld AIM,

$$\mathcal{H}_X^H = S[\rho_X^H] - S[\rho_X^0] = -\int \rho_X^H(\boldsymbol{r}) \log\rho_X^H(\boldsymbol{r})\, d\boldsymbol{r} + \int \rho_X^0(\boldsymbol{r}) \log\rho_X^0(\boldsymbol{r})\, d\boldsymbol{r}$$

$$\equiv \int h_X^H(\boldsymbol{r})\, d\boldsymbol{r}, \tag{174}$$

for selected linear molecules of Table 1 are listed in Table 3 together with the corresponding stockholder- and free-atom values of the Shannon entropy. The preceding equation also defines the atomic entropy-displacement density $h_X^H(\boldsymbol{r})$, shown in Figures 10 and 11.

A reference to H_2 and N_2 entries of Tables 1 and 3 shows that the atomic entropy displacements for bonded hydrogen atoms are approximately additive: $2\mathcal{H}_X^H \cong \mathcal{H}[\rho(H_2)]$. A similar near-additivity is observed for most of the remaining molecules, with the largest deviation from such an uncoupled (independent) behavior of changes in the AIM entropy being observed for the most ionic Li–F bond exhibiting the largest amount of CT.

In general, the bonded atom exhibits a lower degree of uncertainty compared to the free-atom value, a clear sign of the dominating effect of a relatively more compact electron distribution in a molecule. In strong electron acceptors, e.g., for F in LiF and N in HCN, one detects positive displacements due to the dominating CT contribution, which should result in a softer atomic distribution of electrons, thus exhibiting more "disorder" (uncertainty).

The atomic entropy displacements for HF and LiF indicate that the *donor* atom exhibits the dominating (negative) displacement, while the *acceptor* AIM only slightly increases its entropy (see also the net AIM charges in Table 2). The triatomic data in Table 3 provide an additional confirmation of this rule, with an exception of N in HNC. A reference to the atomic charges reported in Table 2 again shows a strong sensitivity of the atomic entropy displacements to the magnitude of the atomic net charge transfer. As expected, a degree of this sensitivity to a change in the atomic overall electron population decreases with a growing overall number of electrons on the atom in question. Indeed, a given displacement in the AIM charge is seen to produce relatively larger reconstructions of the free-atom electron distributions in H or Li, compared to N or F.

In Figure 10 we have compared the contour maps of the density-difference function for the Hirshfeld AIM,

$$\Delta\rho_X^H(\mathbf{r}) = \rho_X^H(\mathbf{r}) - \rho_X^0(\mathbf{r}), \quad (175)$$

with the corresponding entropy displacement density $h_X^H(\mathbf{r})$, for the constituent AIM of the representative diatomics of Figure 2. The density-difference plot (left panel) for the "stockholder" hydrogen in H_2, $H[H_2]$, exhibits typical changes for the covalently-bonded atom, which have been already observed in the density profile of Figure 8: the electron density buildup around the nucleus and in the bond region between nuclei, at the expense of the outer, *non*-bonding regions of the atomic density distribution. This is due to the contraction of the bonded hydrogen and its polarization towards the other atom. This cylindrical, directional polarization of the atomic electron density is clearly seen in the left, electron-density panel. A similar pattern, with a somewhat more emphasized bond polarization, is seen in the entropy-difference plot of the right contour map. The bonding part of the AIM entropy displacement is positive, thus marking an increase in the local uncertainty due to the electron delocalization towards the bonding partner.

A reference to part *b* of the figure, devoted to $N[N_2]$, again shows a general similarity between the two compared contour diagrams. The displacements observed in the valence shell of both panels accord with the known density-deformation changes in the triply-bonded nitrogen: the $(2s,2p_\sigma)$-hybridization along the molecular axis, a transfer of the $2p_\pi$ electrons to the bond-charge region between two nuclei, with the buildup (lowering) of the electron density of the left panel giving rise to the associated increase (decrease) in the local electron uncertainty of the right panel. Both these diagrams confirm the molecular character of the Hirshfels atoms, with the atom displacement "tails" extending all over the molecule.

Table 3. Displacements of the Hirshfeld AIM entropies (in bits) for representative linear molecules of Figure 2. The molecular and promolecular entropy data are also reported. The corresponding molecular entropy shifts are reported in Table 1

Molecule	X	$\mathcal{H}_X^H$	$S[\rho_X^H]$	$S[\rho_X^0]$
H_2	H	− 0.41	3.77	4.18
N_2	N	− 0.34	5.86	6.20
HF	H	− 1.09	3.09	4.18
	F	0.03	1.22	1.19
LiF	Li	− 4.02	3.87	7.89
	F	0.97	2.14	1.17
HCN	H	− 0.87	3.31	4.18
	C	− 0.73	7.29	8.03
	N	0.15	6.35	6.20
CNH	C	0.01	8.04	8.03
	N	− 0.44	5.76	6.20
	H	− 0.98	3.20	4.18

In both density and entropy panels for H[HF], again strongly resembling one another, the hydrogen is seen to be strongly polarized towards the fluorine atom, with the transferred electron density being channeled to both the σ bond region and to the $2p_\pi$ lone-pair regions in fluorine. The valence-shell part of the F[HF] panels reveals a similar polarization of the fluorine towards the hydrogen, with the accompanying increase in the lone pair ($2p_\pi$) density, in the direction perpendicular to the bond axis. These observations confirm a relatively strong covalency of H—F bond. The localization of the fluorine bond-charge close to the proton position provides an additional support to this observation.

Thus, the observed features of the local entropy displacement indicate that it can indeed serve as an alternative, sensitive diagnostic tool for monitoring the bonding condition of AIM.

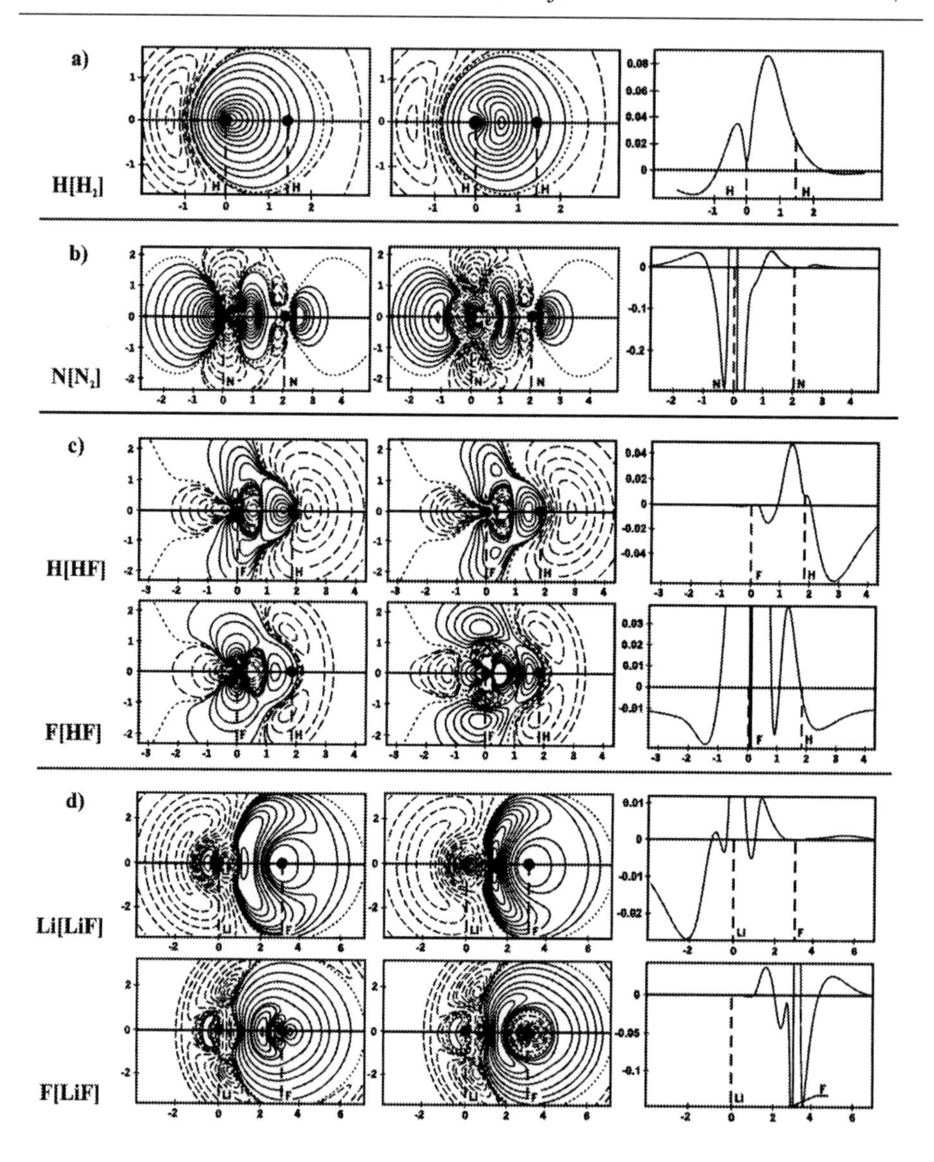

Figure 10. Representative contour maps of the electron density-difference function of bonded stockholder atoms, $\Delta\rho_X^H(\boldsymbol{r})$ (first column), and the associated entropy-displacement density $h_X^H(\boldsymbol{r})$ (second column), for the constituent atoms of diatomic molecules shown in Figure 2: H_2 (*a*), N_2 (*b*), HF (*c*), and LiF (*d*). The contour values are not equidistant, having been selected only for the purpose of revealing the topographic features of the quantities compared. In the third column the bond axis profiles of $h_X^H(\boldsymbol{r})$ (a.u.) are reported [7,10]

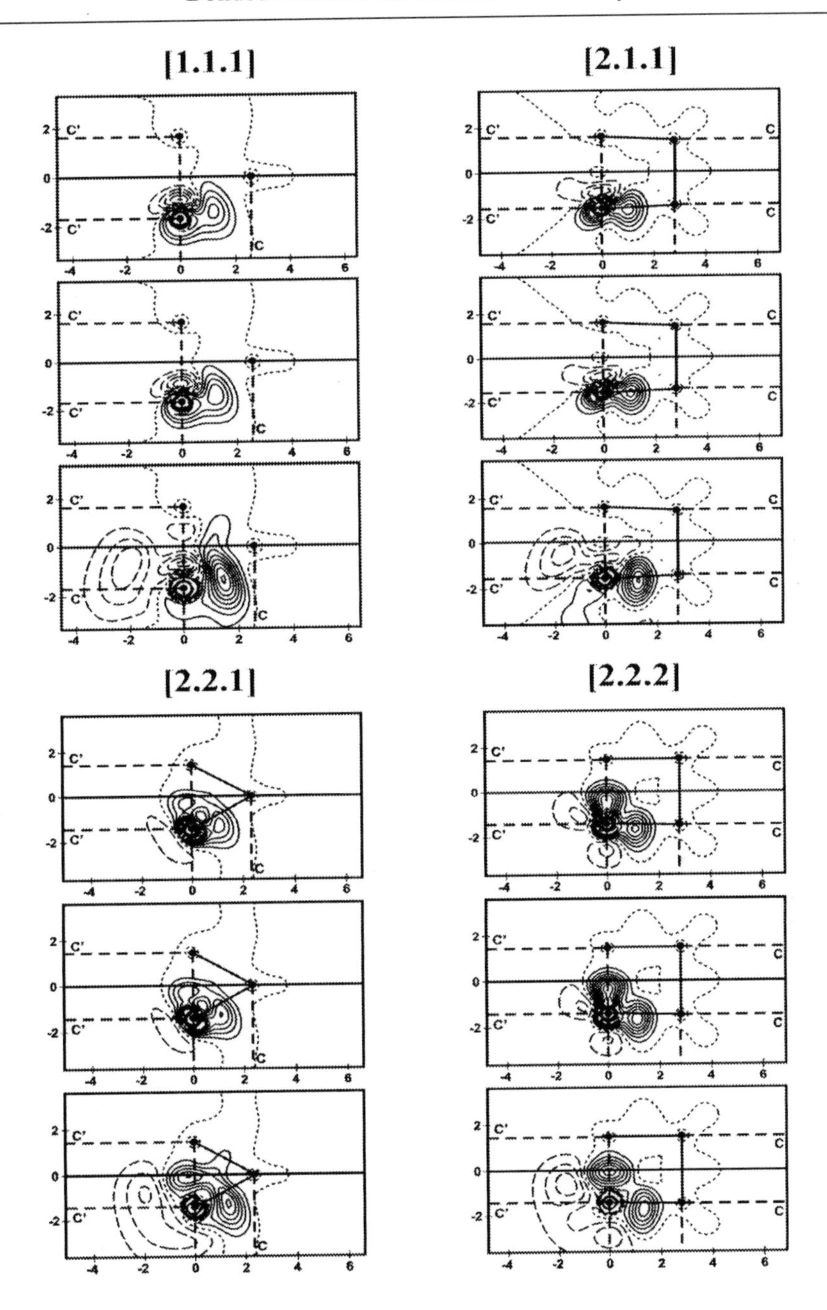

Figure 11. Contour diagrams of displacements in the electron density (upper panel), information-distance density (medium panel) and of the atomic Shannon entropy (lowest panel), for the stockholder bridgehead carbon atoms in the four propellanes of Figure 5

A different CT pattern is found in the strongly ionic Li—F bond, for which the difference diagrams of the AIM density and entropy consistently show a transfer of electrons from the peripheral part of the Li[LiF] electron distribution toward F as a whole, giving rise to a much lower degree of localization of the density displacement. This reflects a partially covalent bond character, with much lower covalent (electron-sharing) contribution though, compared to that observed in HF. This observation accords with the relative hardnesses of the two atoms in these molecules: the two hard atoms in HF give rise to strongly covalent bond, while the soft (Li) and hard (F) atoms in LiF generate a relatively more ionic (CT) bond.

As a final illustration of this section we examine the bridgehead carbon atoms of the four propellanes of Figure 5. The contour maps of Figure 11 report the density difference function, $\Delta\rho_X^H$, the KL missing-information density, Δs_X^H, and the entropy-displacement density, h_X^H. They fully support the conclusions already drawn from the molecular plots of Figures 6 and 7.

The three contour diagrams for the given bridgehead carbon atom in Figure 11 are seen to be qualitatively similar, thus further validating the usefulness of all these quantities as sensitive probes into changes the free atoms undergo in the molecule. These alternative diagnostic tools are thus seen to be to a large extent equivalent.

For the small [1.1.1] and [2.1.1] propellanes, which are lacking the "through-space" component of the central chemical bond, between the bridgehead carbons, the atomic density/information buildups observed in the three panels indeed reflect the "through-bridge" chemical bonds only, with a distinct lowering of the AIM densities in the direction of the other bridgehead atom. In two larger systems of [2.2.1] and [2.2.2] propellanes, in which the presence of the "through-space" component of the central bond has been inferred from both the molecular density difference and the direct bond-order measures, the displacement in AIM electron density and the corresponding information-distance/entropy densities all exhibit increases in directions of both the neighboring bridgehead and bridge carbon atoms.

Several attractive features of the stockholder atomic pieces of the molecular electron distribution make them attractive concepts for chemical applications [7]. In particular, these entropy-equilibrium pieces of the molecular electron density were shown to give rise to the local inter-fragment equalization of the information-distance densities. The role of the entropy-penalty term in the variational principle of the entropy-deficiency in the atomic resolution, which is responsible for the Hirshfeld density localization around the atomic nucleus, has been examined and the additivity of the information distances of Hirshfeld AIM has been established. In addition to the variational principle of the extremum *additive* entropy-deficiency of atomic components of the molecular electron density (see Section 5.3) the complementary variational rule has been formulated for the extremum of the *non-additive* missing-information in atomic densities relative to the free-atom distributions [7,19]. The latter has been shown to also generate the stockholder partition of the molecular electron distribution into atomic fragments.

Displacements of the electron density of bonded atoms from the corresponding free-atom references and the associated missing-information densities and changes in the Shannon entropy density, have all been used to diagnose the atomic promotion to their valence-states in molecules. All these probes were shown to give a consistent diagnosis of the main changes the atoms undergo, when they form the chemical bonds. These novel entropic concepts provide the atomic density-difference function the complementary information interpretation. The reported illustrative applications demonstrate the potential of IT in extracting the chemical interpretation from the known molecular electron distributions and reflect the information origins of the chemical bonds.

It has also been argued that the Hirshfeld atoms equalize their chemical potentials and are in principle the (effective external potential)–representable. This adds the physical meaning to these density-constructs and introduces an element of causality in treating their displacements in a changing molecular environment. The *charge sensitivities* (CS) of the Hirshfeld atoms, e.g., their softness and Fukui Function descriotors, have been expressed in terms of the share-factors of atomic fragments as fractions of the corresponding molecular properties. An overview of the *chemical softness* (density-response) and *hardness* (potential-response) properties of the Hirshfeld atoms has been given in ref. [7].

Chapter 5

IMPORTANCE OF NON-ADDITIVE INFORMATION MEASURES

5.1. ALTERNATIVE RESOLUTIONS AND THEIR ADDITIVE AND NON-ADDITIVE COMPONENTS

Each scheme for resolving the molecular ground-state electron density/probability distribution, e.g., $\rho = \sum_\alpha \rho_\alpha$, into the corresponding molecular freagment, AIM, MO or AO pieces $\boldsymbol{\rho} = \{\rho_\alpha\}$, implies the associated division of the molecular physical quantity $A \equiv A[\rho]$, the functional of the ground state density ρ represented by the overall (*total*) functional $A^{total}[\boldsymbol{\rho}] \equiv A[\rho]$ into the associated *additive*, $A^{add.}[\boldsymbol{\rho}]$, and *non-additive*, $A^{nadd.}[\boldsymbol{\rho}]$, contributions:

$$A[\rho] \equiv A^{total}[\boldsymbol{\rho}] = A^{add.}[\boldsymbol{\rho}] + A^{nadd.}[\boldsymbol{\rho}], \qquad A^{add.}[\boldsymbol{\rho}] = \sum_\alpha A[\rho_\alpha], \tag{176}$$

and hence:

$$A^{nadd.}[\boldsymbol{\rho}] = A^{total}[\boldsymbol{\rho}] - A^{add.}[\boldsymbol{\rho}] = A[\rho] - \sum_\alpha A[\rho_\alpha]. \tag{177}$$

We have indicated above that in the underlying resolved, "multi-component" system $A[\rho]$ becomes the functional of the whole vector of the subsystem densities: $A[\rho] = A^{total}[\boldsymbol{\rho}]$ [56].

For example, this Gordon-Kim-type division [57] of the kinetic energy functional defines the non-additive contribution, which constitutes the basis of the DFT-embedding concept of Cortona and Wesołowski [58-60]. Such division can be also used to partition the information quantities themselves [7,19,21,25,26]. In particular, the inverse of the non-additive Fisher information in the MO resolution has been shown to define the IT-ELF concept [26], in the spirit of the original Becke and Edgecombe formulation [27], while the related quantity in the AO resolution of the SCF MO theory offers the key CG criterion for localization of chemical bonds in molecular systems [19,24,25].

As an illustration consider again the *local* variational problem of Eq. (167), in partitioning the molecular electron density into the AIM components, in which the molecular probability density at point $\boldsymbol{r}$ is divided into atomic contributions in accordance with the conditional probabilities (*share* factors) $\boldsymbol{d}(\boldsymbol{r}) = \{P(X|\boldsymbol{r}) = p_X(\boldsymbol{r})/p(\boldsymbol{r}) = \rho_X(\boldsymbol{r})/\rho(\boldsymbol{r})\}$. The promolecular reference similarly determines the initial conditional probabilities of the free atoms: $\boldsymbol{d}^0(\boldsymbol{r}) = \{P^0(X|\boldsymbol{r}) = p_X{}^0(\boldsymbol{r})/p^0(\boldsymbol{r}) = \rho_X{}^0(\boldsymbol{r})/\rho^0(\boldsymbol{r})\}$. In this local scheme the overall distributions $p(\boldsymbol{r})$ and $p^0(\boldsymbol{r})$ as well as the promolecular conditional probabilities $\boldsymbol{d}^0(\boldsymbol{r})$ are fixed. Therefore, only the molecular conditional probabilities of AIM, $\boldsymbol{d}(\boldsymbol{r})$, are to be determined from the corresponding (entropic) Euler-Lagrane equations, which define the optimum shapes of bonded atoms. The optimized information density $\Delta S[\boldsymbol{d}(\boldsymbol{r})|\boldsymbol{d}^0(\boldsymbol{r})]$ in this local partition is additive in AIM contributions,

$$\Delta S[\boldsymbol{d}(\boldsymbol{r})|\boldsymbol{d}^0(\boldsymbol{r})] = \Delta S^{add.}[\boldsymbol{d}(\boldsymbol{r})|\boldsymbol{d}^0(\boldsymbol{r})] = \sum_X \Delta S_X[P\,(X|\boldsymbol{r})|P^0(X|\boldsymbol{r})],$$

$$\Delta S_X[P(X|\boldsymbol{r})|P^0(X|\boldsymbol{r})] = P(X|\boldsymbol{r}) \log[P(X|\boldsymbol{r})/P^0(X|\boldsymbol{r})], \tag{178}$$

and for the optimum solutions $\Delta S^{nadd.}[\boldsymbol{d}^H(\boldsymbol{r})|\boldsymbol{d}^0(\boldsymbol{r})] = 0$. Thus, the non-additive entropy deficiency reaches the (absolute) minimum value $\Delta S^{add.}[\boldsymbol{d}^H(\boldsymbol{r})|\boldsymbol{d}^0(\boldsymbol{r})] = 0$, for the Hirshfeld (stockholder) solutions, when $\boldsymbol{d}^H(\boldsymbol{r}) = \boldsymbol{d}^0(\boldsymbol{r})$.

The same feature is observed in the KL information densities of the (molecularly-normalized) probability distributions $\boldsymbol{p}^H(\boldsymbol{r}) = \{p_X{}^H(\boldsymbol{r})\}$, relative to those characterizing the free atoms: $\boldsymbol{p}^0(\boldsymbol{r}) = \{p_X{}^0(\boldsymbol{r})\}$. One of the remarkable features of the stockholder division is that the density of the overall entropy deficiency between the molecular and promolecular electron probability densities, which determines the *total* missing-information density in the stockholder-AIM resolution,

$$\Delta S[p(\boldsymbol{r})|p^0(\boldsymbol{r})] = p(\boldsymbol{r}) \log[p(\boldsymbol{r})/p^0(\boldsymbol{r})] = p(\boldsymbol{r})\, I(\boldsymbol{r}) = p(\boldsymbol{r})\, I_X{}^H(\boldsymbol{r})$$

$$\equiv \Delta S^{total}[\boldsymbol{p}^H(\boldsymbol{r})|\boldsymbol{p}^0(\boldsymbol{r})] = \Delta S^{add.}[\boldsymbol{p}^H(\boldsymbol{r})|\boldsymbol{p}^0(\boldsymbol{r})] + \Delta S^{nadd.}[\boldsymbol{p}^H(\boldsymbol{r})|\boldsymbol{p}^0(\boldsymbol{r})], \tag{179}$$

is exactly equal to its *additive* part:

$$\Delta S[p(\boldsymbol{r})|p^0(\boldsymbol{r})] = \Sigma_X p_X{}^H(\boldsymbol{r})\, I_X{}^H(\boldsymbol{r})$$

$$= \Sigma_X \Delta S_X[p_X(\boldsymbol{r})|p_X{}^0(\boldsymbol{r})] \equiv \Delta S^{add.}[\boldsymbol{p}^H(\boldsymbol{r})|\boldsymbol{p}^0(\boldsymbol{r})], \quad (180)$$

due to the common (molecular) surprisal for all stockholder AIM. In other words, the equality of the preceding equation implies that only for this particular partitioning the *non-additive* information-distance function between $\boldsymbol{p}^H(\boldsymbol{r})$ and $\boldsymbol{p}^0(\boldsymbol{r})$ vanishes identically, $\Delta S^{nadd.}[\boldsymbol{p}^H(\boldsymbol{r})|\boldsymbol{p}^0(\boldsymbol{r})] = 0$, and hence also functional $\Delta S^{nadd.}[\boldsymbol{p}^H|\boldsymbol{p}^0] = 0$. The normalization-constrained minimum principle of $\Delta S^{add.}[\boldsymbol{p}|\boldsymbol{p}^0]$ thus implies the associated minimum of the (positive) $\Delta S[p|p^0] = \Delta S^{total}[\boldsymbol{p}|\boldsymbol{p}^0]$ at the optimum, stockholder solution.

5.2. Electron Localization Function

Consider now the *non*-additivities of the Fisher measure of information in the MO resolution, say, defined by the occupied KS MO from DFT calculations (see Section 2.5). In the spin-resolved approach one uses the energy functional of spin densities $\{\rho_\sigma(\boldsymbol{r})\}$ [Eq. (92)], for the spin-like electrons occupying the corresponding MO subsets, $\boldsymbol{\psi} = \left\{ \boldsymbol{\psi}_\sigma = \{\psi_i^\sigma\} \right\}$,

$$\rho_\sigma(\boldsymbol{r}) = \sum_{i\in\sigma}^{occd.} \left|\psi_i^\sigma(\boldsymbol{r})\right|^2 . \quad (181)$$

The double kinetic energy density (a.u.),

$$\tau_\sigma[\boldsymbol{r};\rho_\sigma] = \sum_{i\in\sigma}^{occd.} \left|\nabla\psi_i^\sigma(\boldsymbol{r})\right|^2 , \quad (182)$$

is then proportional to the density of the *multi*-component (*additive*) Fisher information functional in the amplitude representation [see Eq. (108)]:

$$I[\boldsymbol{\psi}_\sigma] = I^{add.}[\boldsymbol{\psi}_\sigma] = 4\sum_{i\in\sigma}^{occd.} \int\left|\nabla\psi_i^\sigma(\boldsymbol{r})\right|^2 d\boldsymbol{r} \equiv \int f_\sigma^{add.}(\boldsymbol{r})\, d\boldsymbol{r} , \quad (183)$$

$$\tau_\sigma(\boldsymbol{r}) = \frac{1}{4} f_\sigma^{add.}(\boldsymbol{r}) . \quad (184)$$

The leading term of the Taylor expansion of the spherically averaged (Hartree-Fock) conditional *pair*-probability of finding in distance s from the reference electron of spin σ at position $\boldsymbol{r}$ the other spin-like electron then reads [27]:

$$P_c^{\sigma\sigma}(s|\boldsymbol{r}) = \frac{1}{3} D_\sigma(\boldsymbol{r}) s^2 + \ldots,$$

$$D_\sigma(\boldsymbol{r}) = \tau_\sigma(\boldsymbol{r}) - \frac{|\nabla\rho_\sigma(\boldsymbol{r})|^2}{4\rho_\sigma(\boldsymbol{r})} = -\frac{1}{4}[f_\sigma^{total}(\boldsymbol{r}) - f_\sigma^{add.}(\boldsymbol{r})] = -\frac{1}{4} f_\sigma^{nadd.}(\boldsymbol{r}) \geq 0. \quad (185)$$

We have explicitly indicated above that the key function $D_\sigma(\boldsymbol{r})$ can be identified as being proportional to the negative non-additive component of the Fisher information in this spin-resolved MO resolution [7,19,26]. The appropriately calibrated square of its inverse, which determines ELF [27], has been successfully used as the probe of the electron localization patterns in molecules [61,62].

A somewhat improved behavior is observed in the DFT-tailored simple inverse of this function, known as IT-ELF [7,19,26]. In Figure 12 representative graphs of both functions are presented for Ne, Ar, Kr and Xe. A qualitative behavior of the two curves is seen to be very similar, emphasizing the shell structure of these noble gas atoms. In general, IT-ELF exhibits smaller outer amplitudes and thus a larger spatial extension than the original ELF.

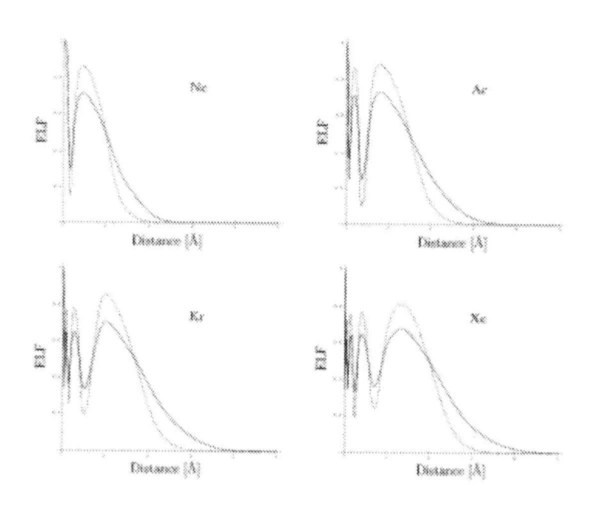

Figure 12. Plots of ELF (dashed line) and IT-ELF (solid line) for Ne, Ar, Kr and Xe

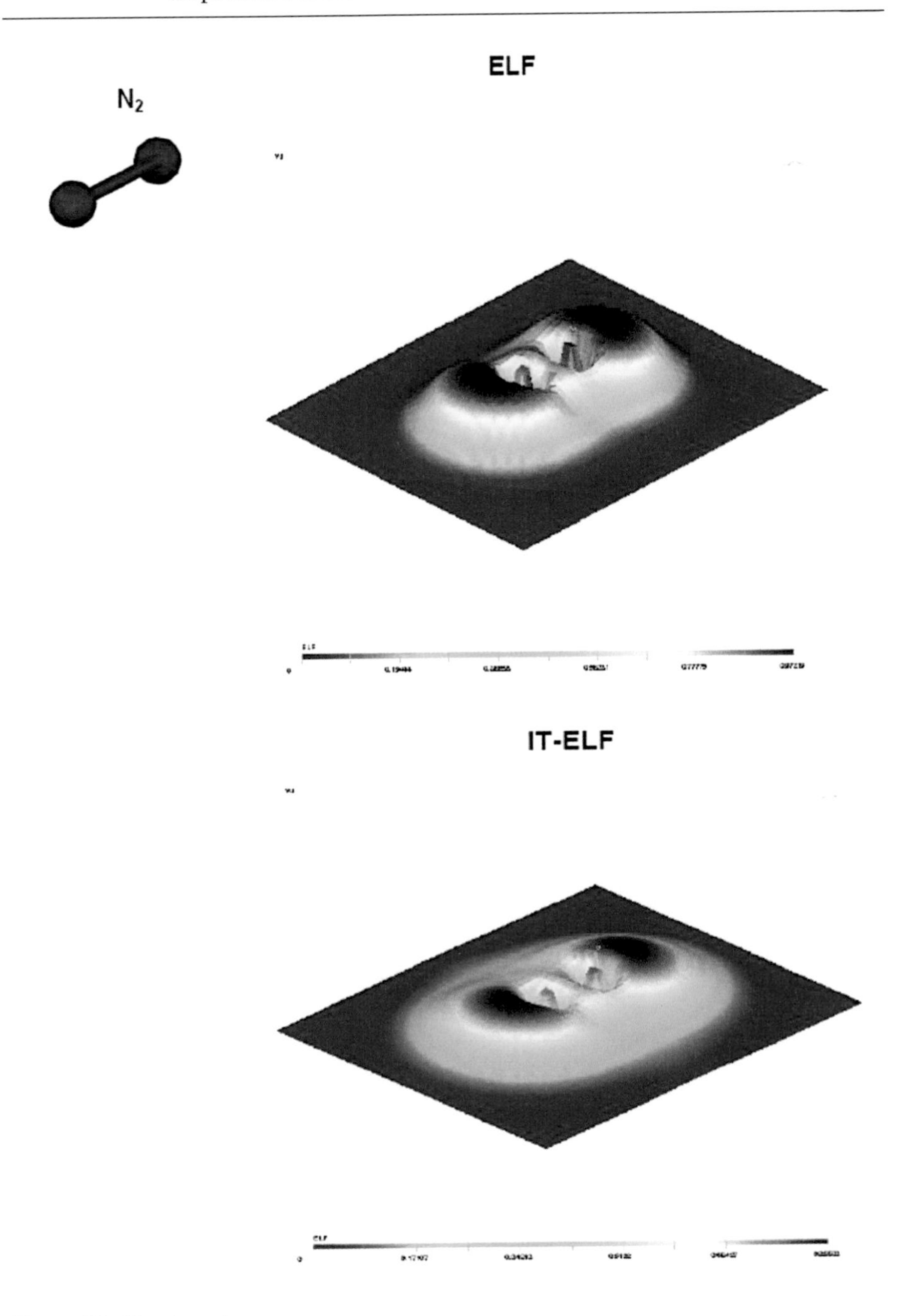

Figure 13. (Continues)

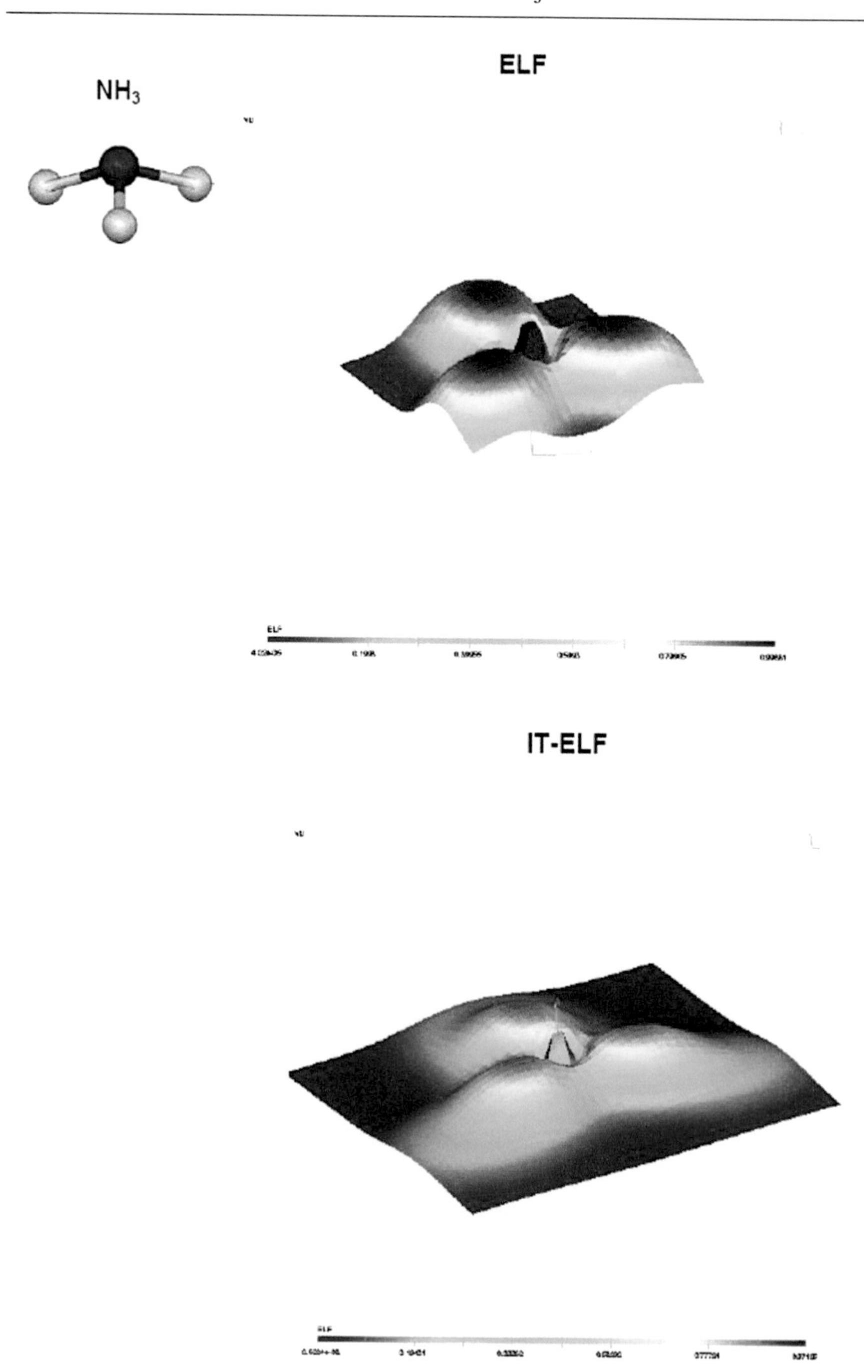

Figure 13. (Continues)

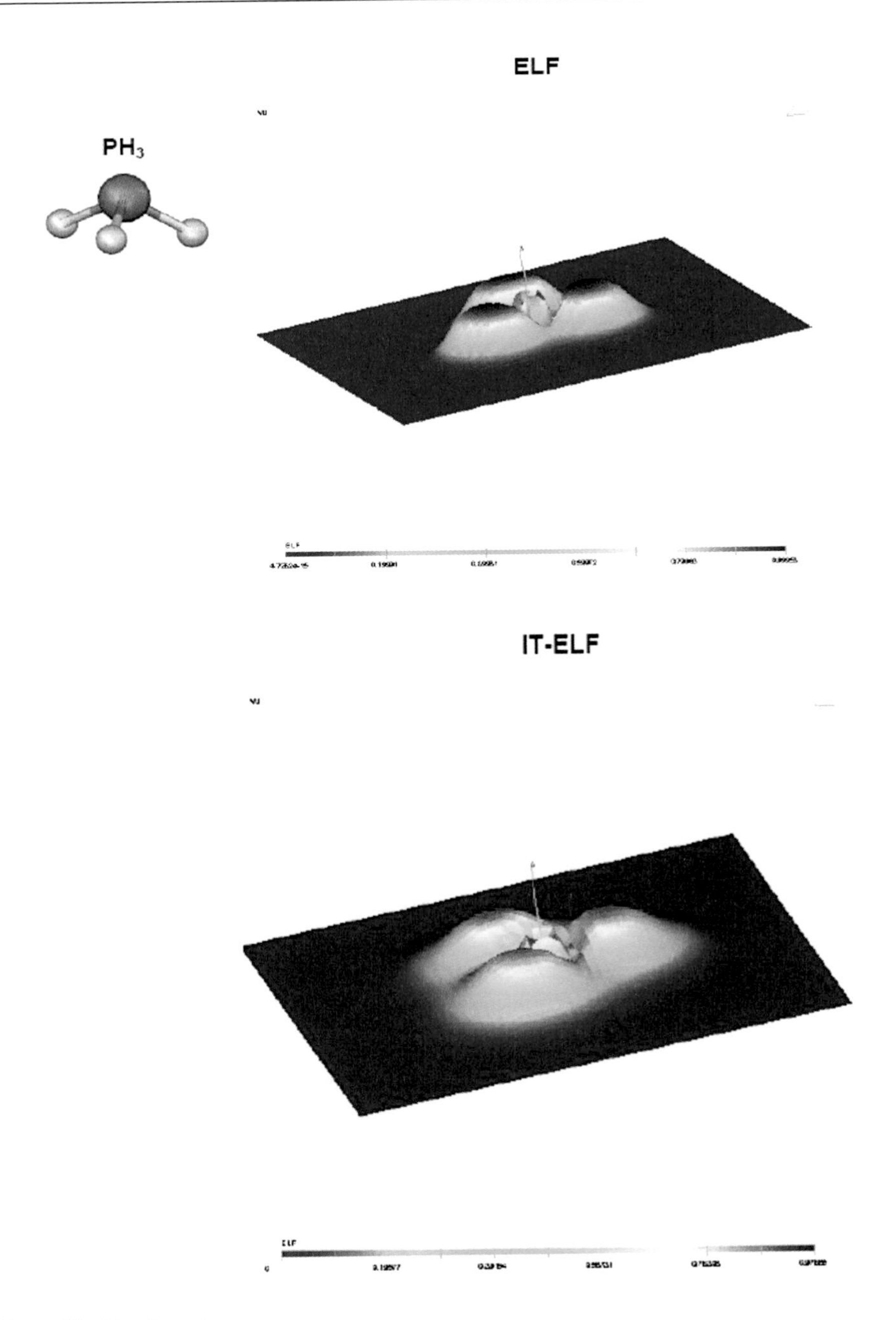

Figure 13. (Continues)

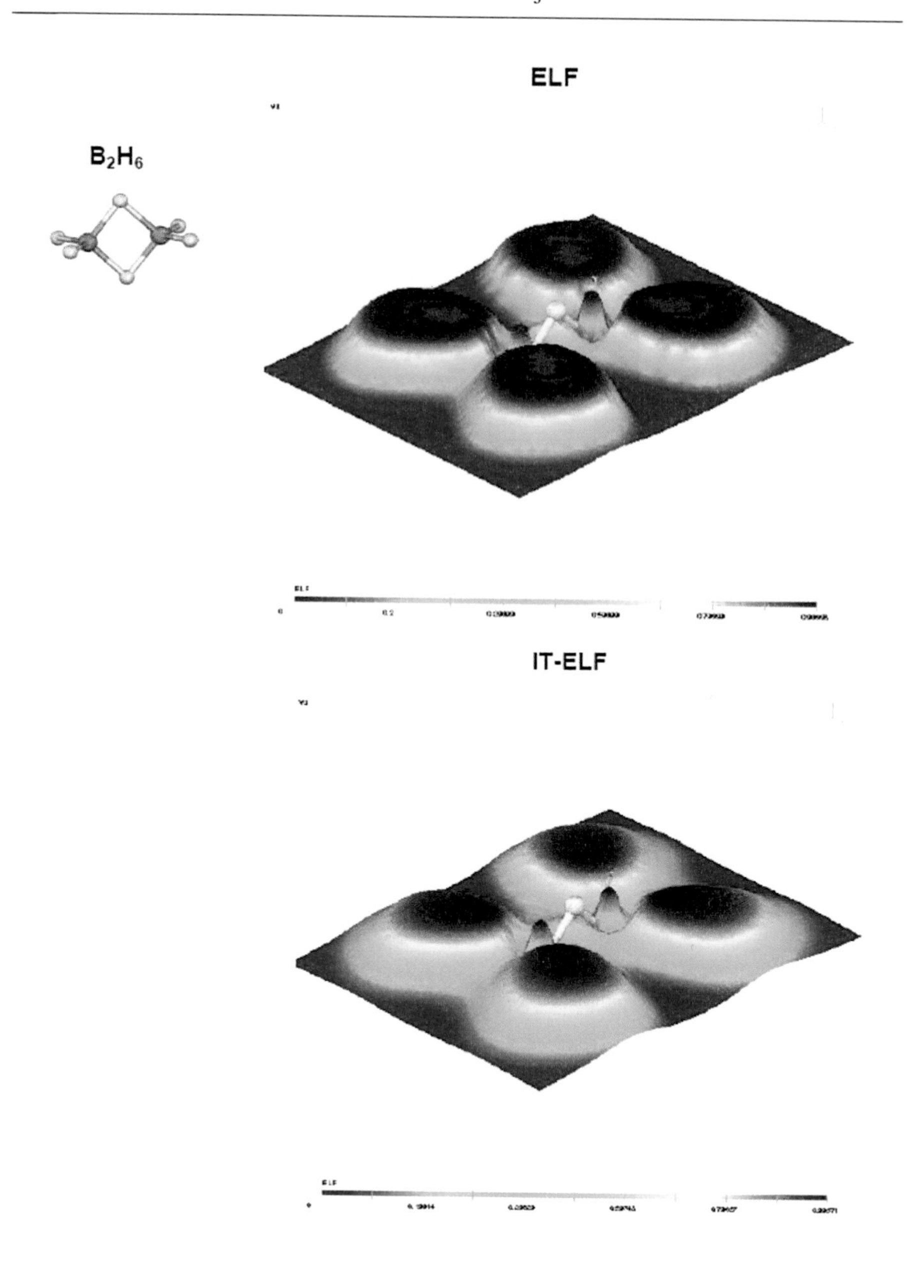

Figure. 13. Comparison between perspective views of the ELF and IT-ELF surfaces for N_2, NH_3, PH_3, and B_2H_6, on the following planes of section: along the bond axis (N_2), on the plane determined by three hydrogen atoms (NH_3 and PH_3), and the plane passing through both terminal BH_2 groups (B_2H_6)

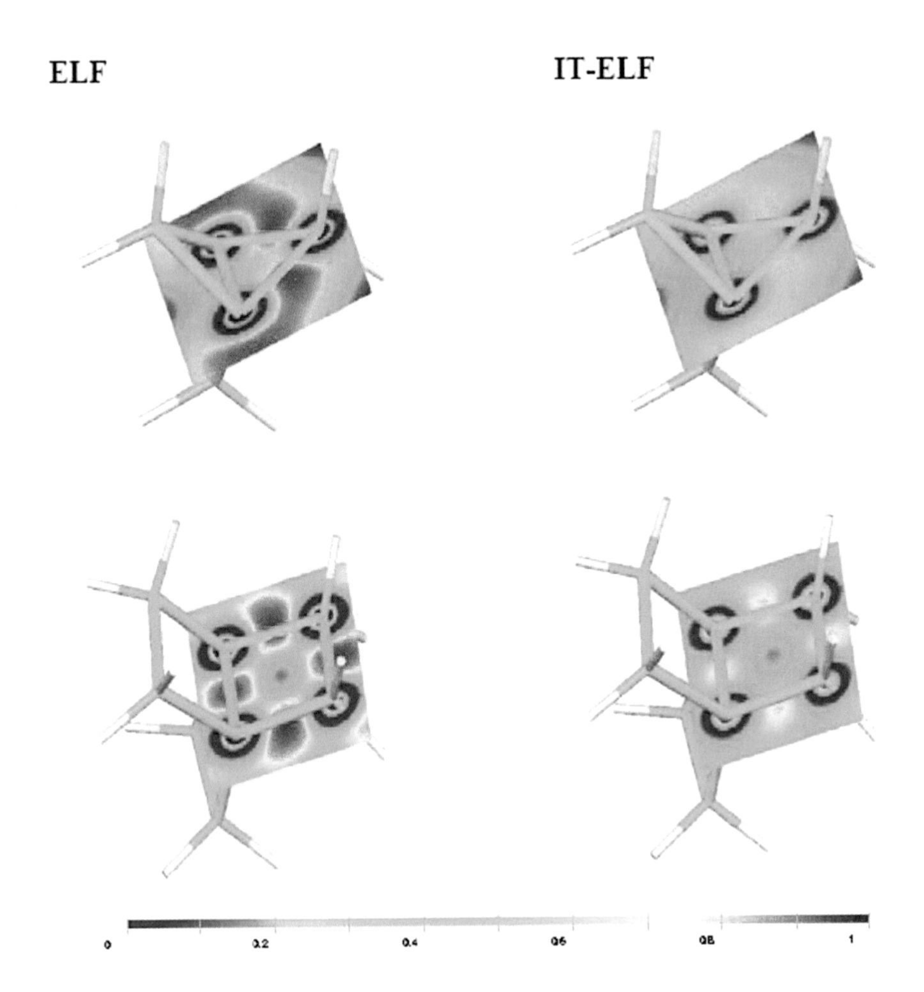

Figure 14. Plots of ELF(first column) and IT-ELF (second column) for the [1,1,1] (top row) and (2,2,2) (bottom row) propellanes of Figure 5 on selected planes of sections. The color scale for the ELF values is given in the bottom of the figure

In Figure 13 we have compared the perspective views of these functions for N_2, NH_3, PH_3 and B_2H_6. These ELF plots convincingly validate the use of this local probe as an indicator of the localization of the valence electrons in the bonding and non-bonding (lone-pair) regions of these illustrative molecular systems. Indeed in the homonuclear diatomic N_2 one detects in both plots the typical accumulation of electrons between the nuclei, due to the formation of the triple covalent bond, and the accompanying increase in the localization of the lone pair electrons in the non-bonding regions of both atoms, a clear manifestation of the accompanying *sp*-hybridization. The three localized N—H bonds are also clearly visualized in both NH_3 panels of the figure and a similar ELF features are detected in its PH_3 part. The final B_2H_6 plots are also seen to successfully locate the bonding electrons of the four terminal B—H bonds.

The final application, shown in Figure 14, again investigates the central and bridge bonds between the carbon atoms in the smallest, [1.1.1] and largest [2.2.2] propellanes of Figure 5. In these contour maps the absence (presence) of the through-space component of the central bond, between the bridgehead carbon atoms, is clearly seen in the upper (lower) panel for both versions of ELF, while the structure of the C—C bonds in the bridges is also transparently revealed. An interesting feature of the bridge bonds in the upper diagrams, also seen in Figure 6, is a slight displacement of the bonding electrons away from the line connecting the bridge and bridgehead carbons, due to the near sp^3-hybridization of the bridge carbon which is required to additionally accommodate the two hydrogen atoms.

5.3. Contra-Gradience Criterion of Bond Localization

Consider next the familiar problem of combining the two Löwdin-orthogonalized AO (OAO), $A(\boldsymbol{r})$ and $B(\boldsymbol{r})$, say, two 1*s* orbitals centered on atoms A and B, respectively, which contribute a single electron each to form the chemical bond A—B. The two basis functions $\chi = (A, B)$ then form the bonding (φ_b) and anti-bonding (φ_a) MO combinations, $\boldsymbol{\varphi} = (\varphi_b, \varphi_a) = \boldsymbol{\chi}\mathbf{C}$:

$$\varphi_b = \sqrt{P}A + \sqrt{Q}B \equiv \textstyle\sum_{k=A,B} \chi_k C_{k,b},$$

$$\varphi_a = -\sqrt{Q}A + \sqrt{P}B \equiv \textstyle\sum_{k=A,B} \chi_k C_{k,a}, \qquad P + Q = 1, \tag{186}$$

where the square matrix $\mathbf{C} = [\boldsymbol{C}_b, \boldsymbol{C}_a]$ groups the LCAO MO expansion coefficients expressed in terms of the complementary probabilities P and $Q = 1 - P$. The former marks the conditional probabilities $P(A|\varphi_b) = P(B|\varphi_a) = P$, and the latter measures the remaining matrix elements of the orbital conditional probabilities of AO in MO: $P(B|\varphi_b) = P(A|\varphi_a) = Q$.

We then examine the (*bonding*) ground-state, $\Psi_0 = [\varphi_b^2]$, the (*non-bonding*) *singly*-excited configuration $\Psi_1 = [\varphi_b^1 \varphi_a^1]$, and the (*anti-bonding*) *doubly*-excited state $\Psi_2 = [\varphi_a^2]$. Consider next the *charge–and–bond-order* (CBO), density matrix of the SCF LCAO MO theory for each of these model configurations,

$$\gamma_i = \mathbf{C}\,\mathbf{n}(\Psi_i)\,\mathbf{C}^{\mathrm{T}} = \left\langle \chi \middle| \psi^{occd.}(\Psi_i) \right\rangle \left\langle \psi^{occd.}(\Psi_i) \middle| \chi \right\rangle, \qquad i = 0, 1, 2, \qquad (187)$$

with the diagonal matrix of the MO electron occupations $\mathbf{n}(\Psi_i) = \{n_s(\Psi_i)\delta_{s,s'}\}$. As shown in the preceding equation, this matrix constitutes the OAO representation of the projection operator on the subspace of the *singly*-occupied spin-MO, $\psi^{occd.}(\Psi_i)$, in the given electron configuration Ψ_i.

The three matrices of Eq. (187) read:

$$\gamma_0 = 2\begin{bmatrix} P & \sqrt{PQ} \\ \sqrt{PQ} & Q \end{bmatrix}, \quad \gamma_1 = \begin{bmatrix} 1 & 0 \\ 0 & 1 \end{bmatrix}, \quad \gamma_2 = 2\begin{bmatrix} Q & -\sqrt{PQ} \\ -\sqrt{PQ} & P \end{bmatrix}. \qquad (188)$$

They are seen to reflect the configuration bonding status. Indeed, the (bonding) ground state exhibits the positive off-diagonal bond-order, $\gamma_{A,B} = 2(PQ)^{1/2}$, which vanishes in the non-bonding configuration Ψ_1, $\gamma_{A,B} = 0$, while its negative value in Ψ_2, $\gamma_{A,B} = -2(PQ)^{1/2}$, correctly reflects the configuration anti-bonding character.

In this (real) 2-OAO model the AO-partition of the Fisher information densities of the bonding and anti-bonding MO gives:

$$f_b = 4(\nabla\varphi_b)^2 \equiv f_\chi^{total}[\varphi_b] = 4[P(\nabla A)^2 + Q(\nabla B)^2] + 8\sqrt{PQ}\,\nabla A\cdot\nabla B$$

$$\equiv f_\chi^{add.}[\varphi_b] + f_\chi^{nadd.}[\varphi_b], \qquad (189)$$

$$f_a = 4(\nabla\varphi_a)^2 \equiv f_\chi^{total}[\varphi_a] = 4[Q(\nabla A)^2 + P(\nabla B)^2] - 8\sqrt{PQ}\,\nabla A\cdot\nabla B$$

$$\equiv f_\chi^{add.}[\varphi_a] + f_\chi^{nadd.}[\varphi_a]. \qquad (190)$$

It should be further realized that along the bond-axis, in the AO-overlap region between the two atoms, which is decisive for the bonding or anti-bonding character of MO (see Figure 15), $\nabla A \cdot \nabla B < 0$, so that for these locations

$$f_\chi^{nadd.}[\varphi_b] = 8\sqrt{PQ}\,\nabla A \cdot \nabla B \equiv 8\sqrt{PQ}\; i^{c-g} < 0,$$

$$f_\chi^{nadd.}[\varphi_a] = -8\sqrt{PQ}\,\nabla A \cdot \nabla B = -8\sqrt{PQ}\; i^{c-g} > 0. \tag{191}$$

In the preceding equation we have introduced the CG density, which in the 2-OAO model is given by the scalar products of gradients of the two interacting orbitals, $i^{c-g} = \nabla A \cdot \nabla B$.

One similarly defines the average densities of the *non*-additive and additive Fisher information contributions per electron for the configuration in question:

$$f_\chi^{nadd.}[\Psi_i] = \Sigma_s\, [n_s(\Psi_i)/N]\, f_\chi^{nadd.}[\varphi_s] \equiv \Sigma_s\, p_s(\Psi_i)\; f_\chi^{nadd.}[\varphi_s],$$

$$f_\chi^{add.}[\Psi_i] = \Sigma_s\, p_s(\Psi_i)\, f_\chi^{add.}[\varphi_s]. \tag{192}$$

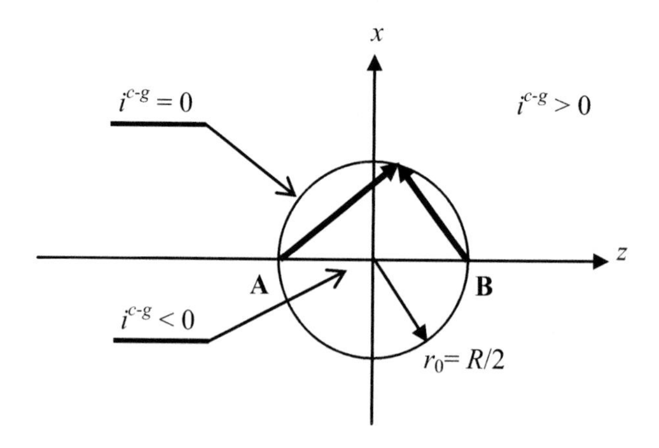

Figure 15. The circular contour of the vanishing CG integrand of two *s* orbitals on atoms A and B, $i^{c-g}(\boldsymbol{r}) = 0$, passing through both nuclei, which separates the *bonding region* (inside the circle), where $i^{c-g}(\boldsymbol{r}) < 0$, from the region of positive contributions $i^{c-g}(\boldsymbol{r}) > 0$ (outside the circle)

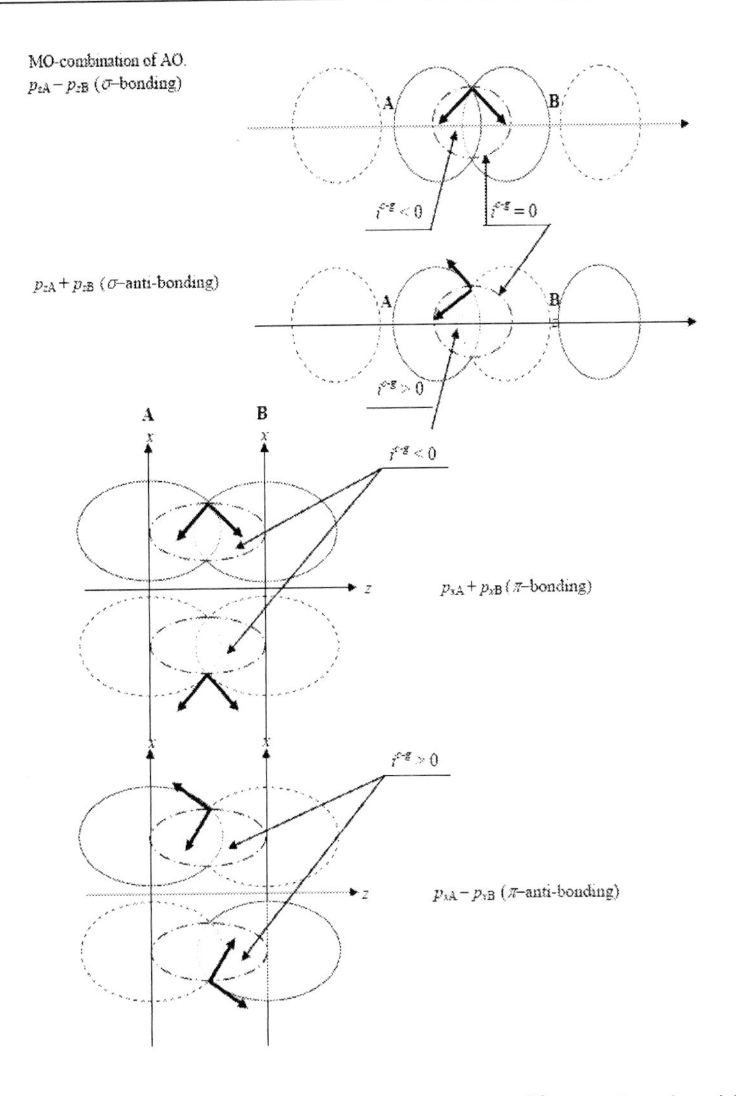

Figure 16. A schematic representation of the bonding [$i^{c\text{-}g}(\boldsymbol{r}) < 0$] and anti-bonding [$i^{c\text{-}g}(\boldsymbol{r}) > 0$] regions of the strong orbital overlap in the chemical interactions between the valence orbitals p_σ and p_π on atoms A and B, enclosed by the vanishing contragradience contours $i^{c\text{-}g}(\boldsymbol{r}) = 0$, represented by the pointed broken line. The AO gradients, each in the direction perpendicular to the orbital contour, are mutually perpendicular on the $i^{c\text{-}g}(\boldsymbol{r}) = 0$ surface (see also Fig. 15). It again follows from these qualitative diagrams that the negative density of the non-additive Fisher information accompanies the electron delocalization in the bonding MO-combination of AO, while the positive density of this information contribution is associated with an effective electron localization, which characterizes the anti-bonding MO-combination of AO

For the three model electron configurations one thus obtains:

$$f_\chi^{nadd.}[\Psi_0] = 8\sqrt{PQ}\, i^{c-g}, \quad f_\chi^{nadd.}[\Psi_1] = 0, \quad f_\chi^{nadd.}[\Psi_2] = -8\sqrt{PQ}\, i^{c-g}. \tag{193}$$

Therefore, the AO-phase dependent, non-additive contributions to the MO Fisher information density, proportional to the CG integral

$$I^{c\text{-}g} = \int i^{c-g}(\boldsymbol{r})\, d\boldsymbol{r} = \int \nabla A(\boldsymbol{r})\cdot\nabla B(\boldsymbol{r})\, d\boldsymbol{r} = -\int A(\boldsymbol{r})\Delta B(\boldsymbol{r})\, d\boldsymbol{r}$$

$$= \frac{2m_e}{\hbar^2}\langle A|\hat{\mathrm{T}}|B\rangle \equiv \frac{2m_e}{\hbar^2} T_{A,B}, \tag{194}$$

reflect the bonding and anti-bonding characters of both MO, as well as the bonding, non-bonding and anti-bonding nature of the three model configurations. As such they may provide attractive concepts for both locating and indexing the chemical bonds in molecules [24,25].

As also indicated in the preceding equation, the CG integral measures the coupling (off-diagonal) element $T_{A,B}$ of the electronic kinetic energy operator $\hat{\mathrm{T}}$. The CG integral is thus proportional to the kinetic-energy matrix element between the two basis functions. Such integrals are routinely calculated in typical quantum-chemical packages for determining the electronic structure of molecular systems. This observation also emphasizes the crucial role of the kinetic energy terms in the IT interpretation of the origins of chemical bonding, which uses the Fisher measure of information.

A reference to Figure 15 indicates that the $i^{c\text{-}g}(\boldsymbol{r}) = 0$ contour is defined by the equation $\boldsymbol{r}_A\cdot\boldsymbol{r}_B = 0$. It separates the region of positive contributions $i^{c\text{-}g}(\boldsymbol{r}) > 0$, outside this contour, from the region of negative CG density, $i^{c\text{-}g}(\boldsymbol{r}) < 0$, inside the contour. Consider its section in xz-plane, for $y = 0$. For the Cartesian reference frame located at the bond mid-point and the two nuclei in positions $\boldsymbol{R}_A = (0, 0, -\frac{1}{2}R)$ and $\boldsymbol{R}_B = (0, 0, \frac{1}{2}R)$, where the internuclear separation $R_{AB} = R$, the electron-position vectors $\{\boldsymbol{r}_X = \boldsymbol{r} - \boldsymbol{R}_X\}$ in this plane of section are: $\boldsymbol{r}_A = (x, 0, z + \frac{1}{2}R)$ and $\boldsymbol{r}_B = (x, 0, z - \frac{1}{2}R)$. The equation determining the $i^{c\text{-}g}(\boldsymbol{r}) = 0$ contour then reads: $x^2 + z^2 = (\frac{1}{2}R)^2 \equiv r_0^2$. It determines the circle centered at the bond mid-point, passing through both nuclei (see Figure 15).

As argued in the qualitative diagram of Figure 16, the negative CG density is also essential for the bonding interaction between two p-orbitals. Therefore, it is essential for the *bonding* chemical interaction between a given pair of basis functions that the gradient of one orbital exhibits a non-vanishing negative component along the direction of the gradient of the other orbital, which justifies the name of the "*contra-gradience*" criterion itself.

5.4. ILLUSTRATIVE APPLICATIONS

In the ground-state of N-electron system and the OAO basis set $\chi = (\chi_1, \chi_2, \ldots, \chi_m)$ the non-additive Fisher information density in the AO resolution for the electron configuration defined by N lowest (*singly*-occupied) molecular *spin-orbitals* $\psi = \{\psi_k\}$, with the spatial MO parts $\varphi = \chi\mathbf{C} = \{\varphi_k,\ k = 1, 2, \ldots, N\}$, e.g., those from the SCF MO or KS calculations, reads:

$$I^{nadd}[\chi] = 4\sum_{k=1}^{m}\sum_{l=1}^{m} \int \gamma_{k,l}(1-\delta_{k,l})\nabla\chi_l^*(\boldsymbol{r})\cdot\nabla\chi_k(\boldsymbol{r})d\boldsymbol{r} \equiv 2\int f^{nadd.}(\boldsymbol{r})d\boldsymbol{r} = 8T^{nadd.}[\chi], \quad (195)$$

where the CBO matrix:

$$\boldsymbol{\gamma} = \langle \chi | \left(\sum_{k=1}^{N} |\psi_k\rangle\langle\psi_k| \right) | \chi \rangle = \langle \chi|\psi\rangle\langle\psi|\chi\rangle = \mathbf{C}\mathbf{C}^{\dagger} = \{\gamma_{u,w}\}. \quad (196)$$

It is proportional to the non-additive component $T^{nadd.}[\chi]$ of the system average kinetic energy: $T[\chi]$ = tr($\boldsymbol{\gamma}\mathbf{T}$), where the kinetic-energy matrix in AO representation $\mathbf{T} = \{T_{k,l} = \langle\chi_k|\hat{\mathrm{T}}|\chi_l\rangle\}$.

In this general molecular scenario of the orbital approximation one uses the most extended (valence) basins of the negative CG density, $f^{nadd.}(\boldsymbol{r}) < 0$, enclosed by the corresponding $f^{nadd.}(\boldsymbol{r}) = 0$ surface, as the location of the chemical bond(s). This proposition has been recently validated numerically [25] using standard SCF MO calculations (GAMESS software) in the minimum Gaussian (STO-3G) basis set. In the remaining part of this section we presents representative results of this extensive study. The contour maps, for the optimized geometries of all molecules, will be reported in a.u. The negative CG basins, also shown in the perspective views, are identified by the broken-line contours. For the visualization purposes the Matpack and DISLIN graphic libraries have been used.

The contour map of Figure 17 confirms the qualitative predictions of Figure 15. In this axial cut of $f^{nadd.}(\boldsymbol{r})$ for H_2 the *non*-additive Fisher information is seen to be lowered in the spherical *bonding* region between the two nuclei. At the same time the accompanying increases in this quantity are observed in the *non-bonding* regions of each hydrogen atom, signifying the increased localization/structure in this homonuclear diatomic due to the axial polarization of the initially spherical atomic densities. It should be stressed, however, that the molecular CG integral over the whole space must be positive, since by the *virial theorem* for the equilibrium geometry the shift in kinetic component of the BO potential, relative to the separated atom (dissociation) limit, must be positive, thus giving rise to the overall *"production"* of the non-additive Fisher information in the molecular hydrogen [19,25].

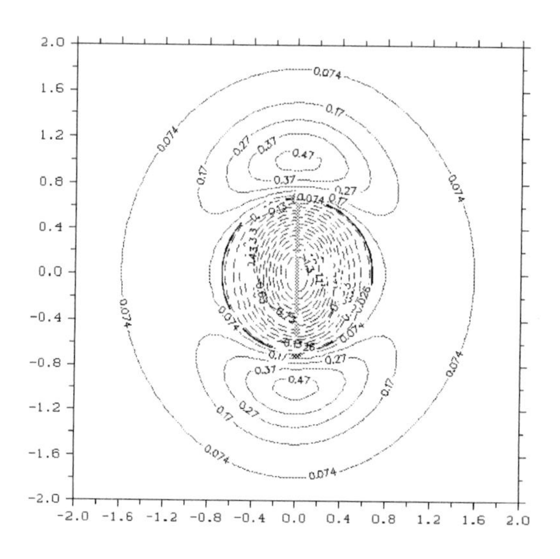

Figure 17. The contour map of the CG density $f^{nadd.}(\boldsymbol{r})$ for H_2 (see also Fig. 15)

A similar analysis for HF is presented in Figure 18. The perspective view of the $f^{nadd.}(\boldsymbol{r}) < 0$ volumes (upper panel) and the contour map of the axial cut of $f^{nadd}(\boldsymbol{r})$ (lower panel) indicate the existence of three basins of a decreased non-additive Fisher information: a large, dominating bonding region Ω_1 located in the valence shells of two atoms, and the axially-centered two small volumes detected in the inner-shell of F. The shape of the bonding volume exhibits its polarization towards the $2p_\pi$ orbitals of a more electronegative fluorine atom.

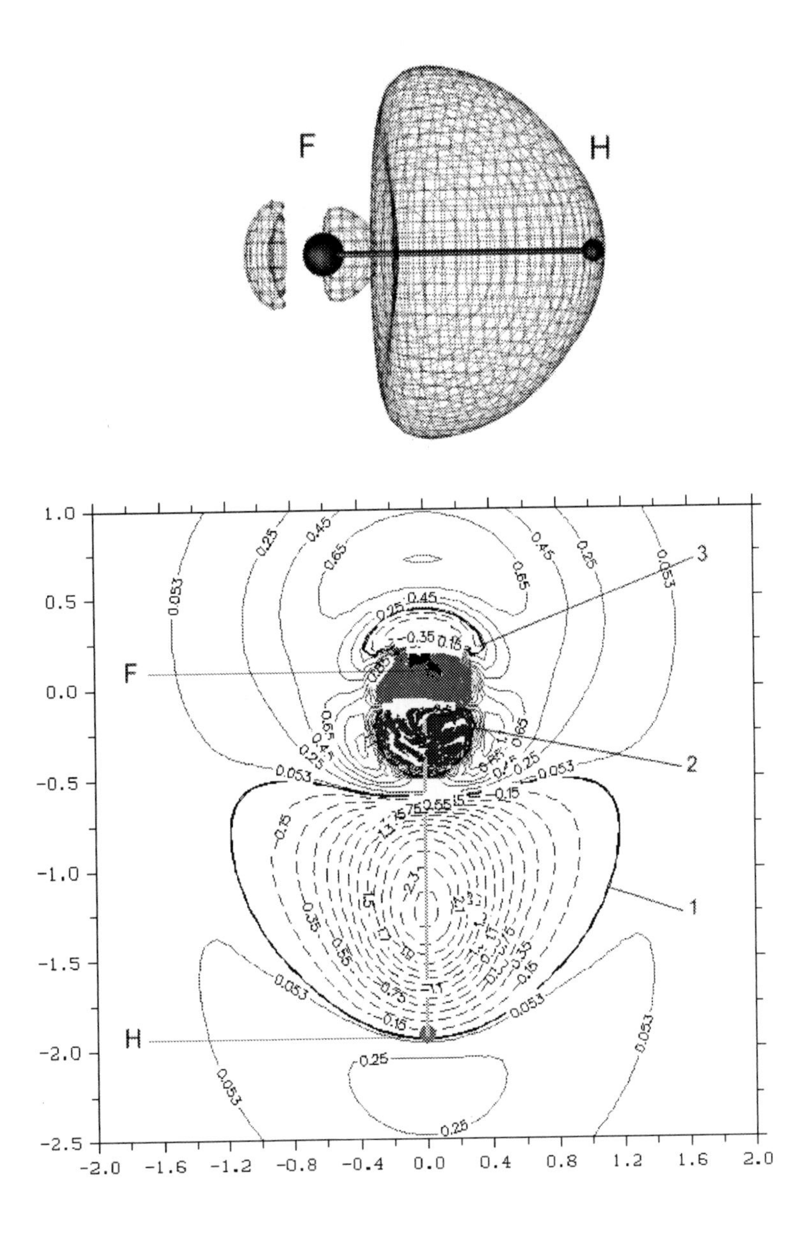

Figure 18. The perspective view of the negative basins of CG (upper panel) and the contour map of the CG density (lower panel) for HF

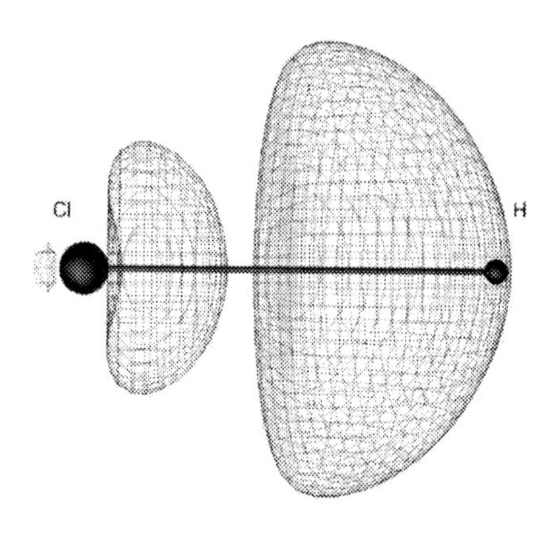

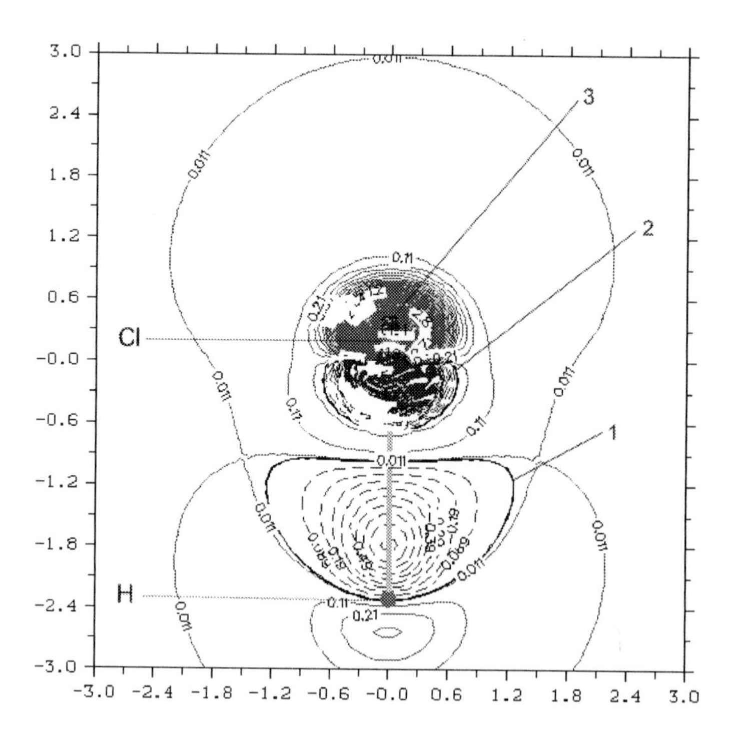

Figure 19. The same as in Figure 18 for HCl

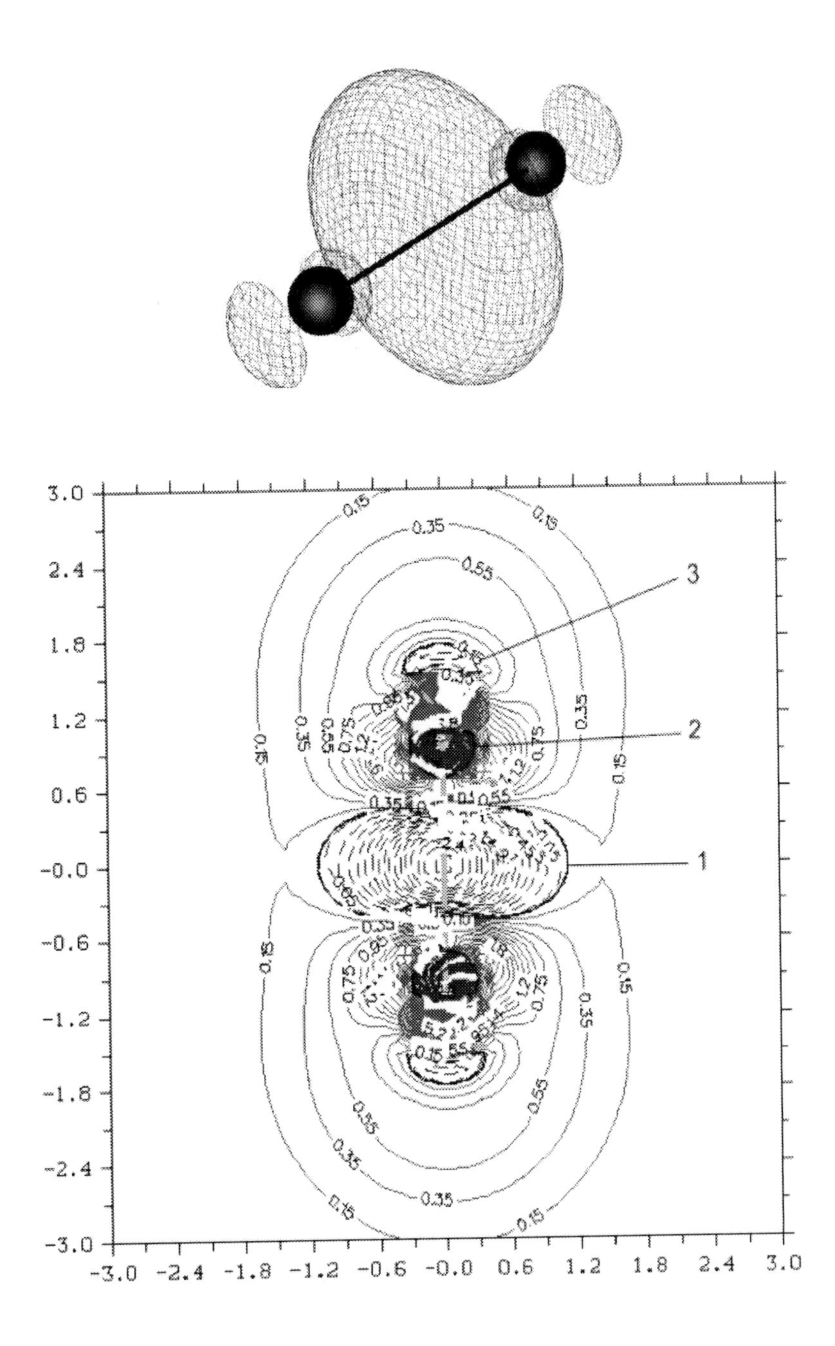

Figure 20. The same as in Figure 18 for N_2

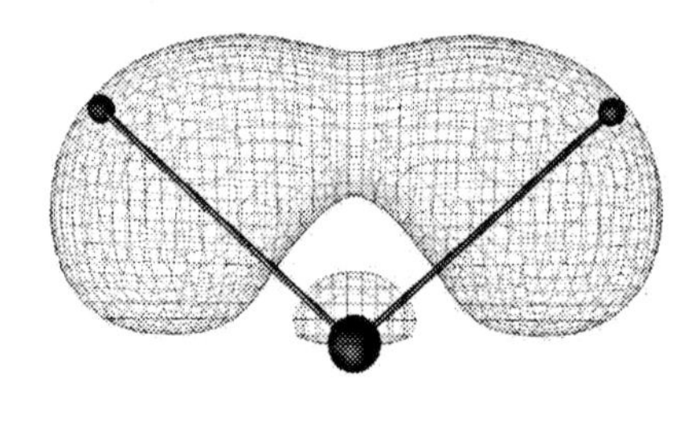

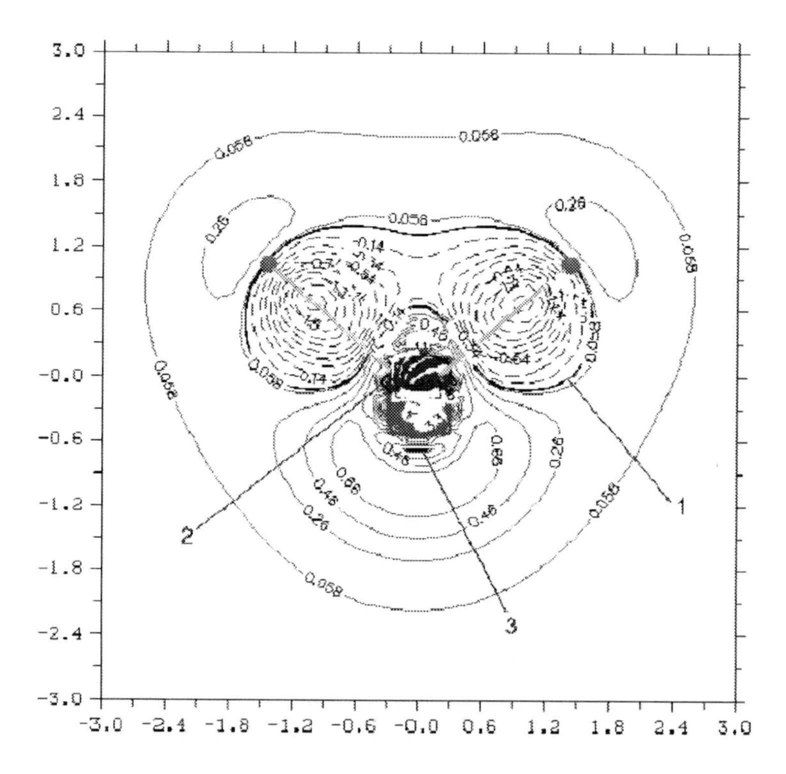

Figure 21. The same as in Figure 18 for H_2O

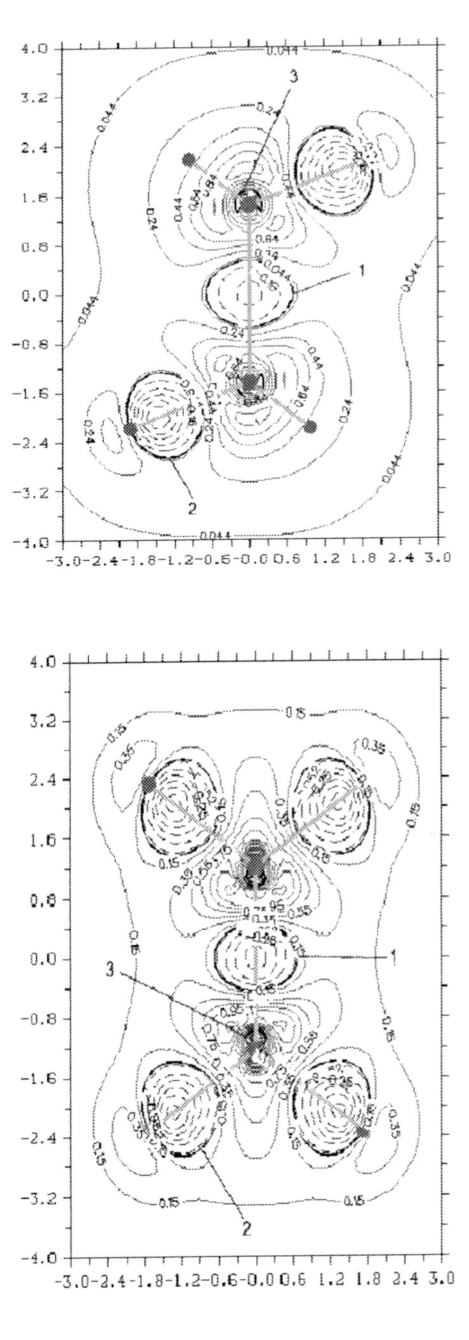

Figure 22. The same as in Figure 17 for ethane (upper panel) and ethylene (lower panel)

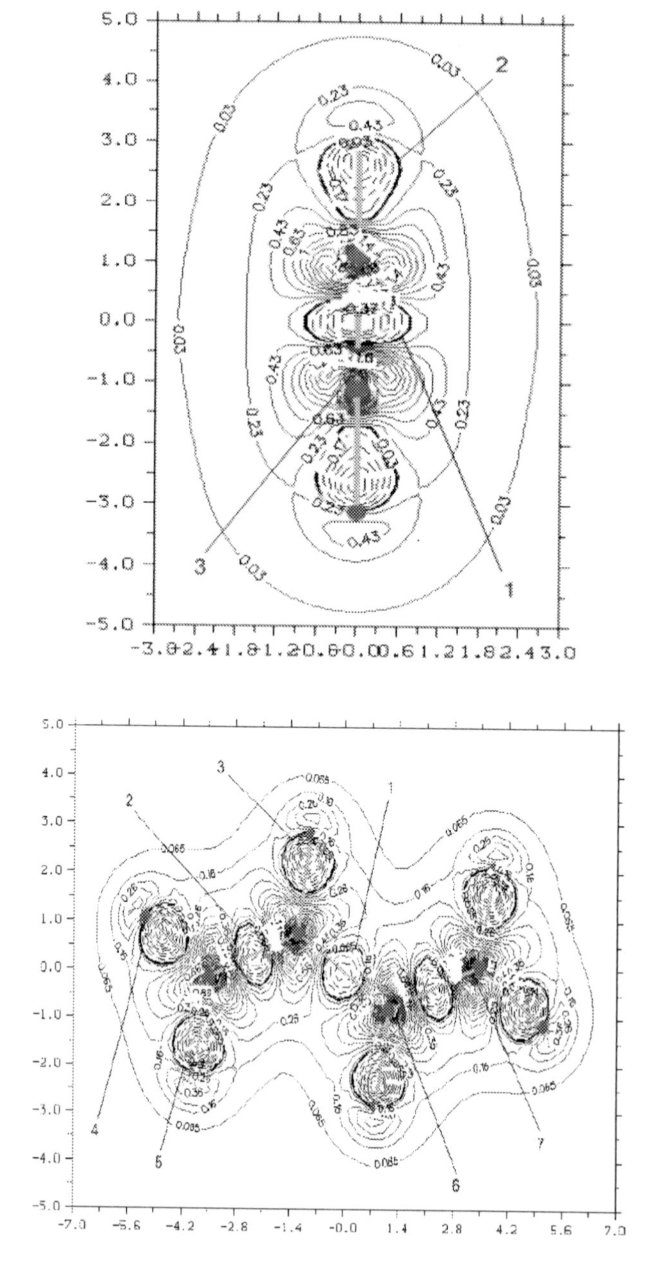

Figure 23. The same as in Figure 17 for acetylene (upper panel) and butadiene (lower panel)

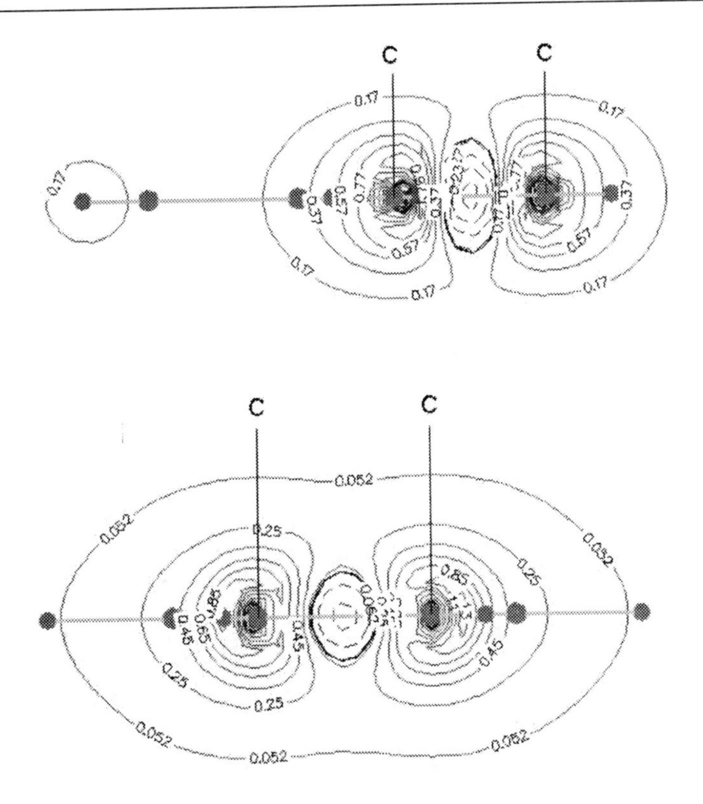

Figure 24. The same as in Fig. 23 for the planes of section perpendicular to the molecular plane in butadiene passing through the peripheral (upper diagram) and middle (lower diagram) C—C bond axis

In HCl (Figure 19) one again observes two smaller (inner-shell) and a large (valence-shell) basins of the negative CG density. The softer heavy atom is now seen to undergo a more substantial inner-shell reconstruction of the non-additive Fisher information. Again, there is an axial build-up of $f^{nadd}(\boldsymbol{r})$ in the non-bonding regions of two atoms, particularly on the hydrogen atom. It should be realized that compared to HF, where both atoms exhibit the "hard" (difficult to polarize) electron distributions in their valence shells, the (soft) chlorine atom combined with the (hard) hydrogen generates a stronger ionic (electron-transfer) component H→Cl, ultimately giving rise to the ionic pair $H^{+}Cl^{-}$ in the dissociation limit, and hence a smaller covalent (electron-sharring) component H—Cl in the resultant chemical bond.

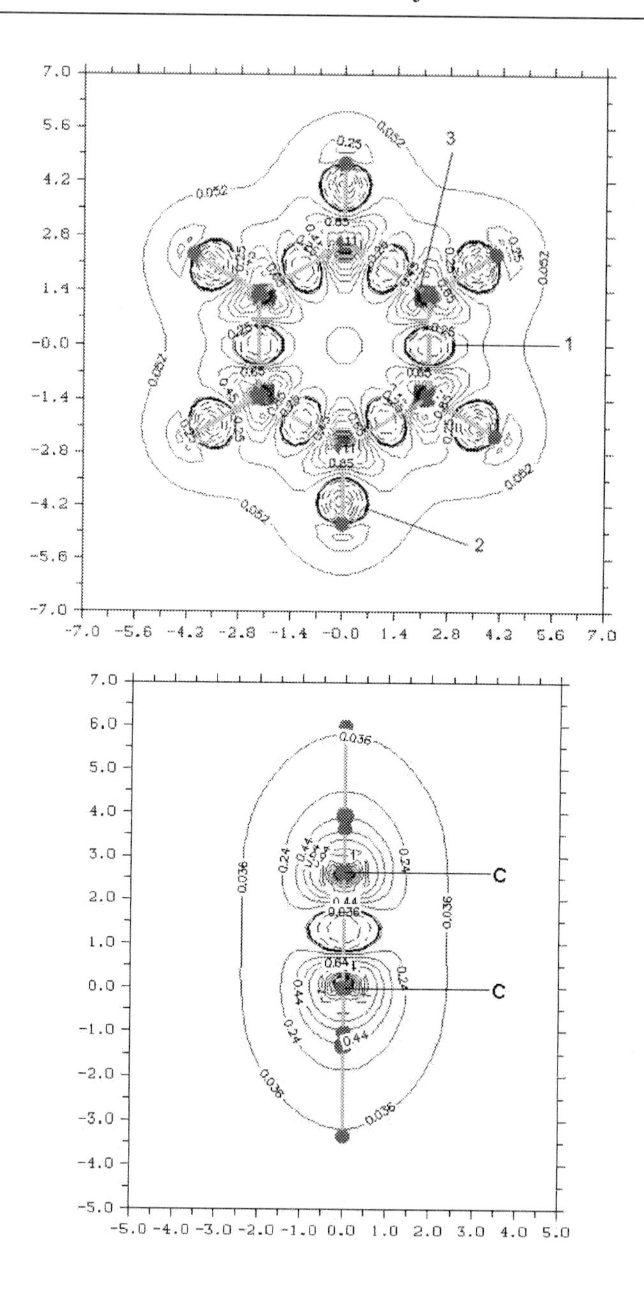

Figure 25. The same as in Figure 17 for benzene. The upper panel shows the contour map in the molecular plane, while the lower panel corresponds to the perpendicular plane of section passing through one of the C—C bonds

Consider next the triple chemical bond in N_2 (Figure 20), where the bonding (valence) basin is now distinctly extended away from the bond axis, due to the presence of two π bonds accompanying the central σ bond. Small core-polarization basins, now symmetrically distributed near each constituent atom along the bond axis, are again observed in the perspective view, while the *sp*-hybridization reconstruction of the non-bonding regions on both atoms is again much in evidence in the accompanying contour map. The dominating (bonding) region around the bond middle-point is now "squeezed" between the two cores of nitrogen atoms. The small, axially placed core regions of the depleted contra-gradience are seen to be surrounded by the volumes of the positive values of this information density in transverse directions. They reflect the charge displacements accompanying the π-bond formation, which is also seen in the corresponding density difference diagram of Figure 2.

One thus concludes on the basis of this numerical evidence that the contra-gradience criterion for detecting the valence basins of a diminished non-additive Fisher information in AO-resolution indeed provides an efficient tool for locating the bonding regions in typical diatomics. It diagnoses all typical displacements of the bonded atoms relative the corresponding free atoms, which accompany the bond formation process, which have already been diagnosed from the density difference diagrams, e.g., the AIM polarization via the promotion/hybridization mechanism, the interatomic CT, and the constructive interference of AO in the bonding region, which is responsible for the electron accumulation between the covalently bonded atoms.

In the water molecule (Figure 21) one detects two, slightly overlapping outer-basins of the negative non-additive Fisher information in the O—H bonding regions, and two small inner-shell basins of the negative CG density on oxygen atom. The bonding basins are located between the corresponding pairs of nuclei, which define the two localized single bonds, and the lowering of the CG density in each bond is seen to be the strongest in the direction linking the two nuclear attractors. The overlapping character of these two regions of the negative non-additive kinetic-energy, reflected by the present non-additive Fisher-information probe, indicates the delocalization of the bonding electrons of one O—H bond into the bonding region of the other chemical bond, as indeed implied by the delocalized character of the occupied canonical MO. The contour map for the cut in the molecular plane also reveals a strong buildup of this information/kinetic-energy quantity in the lone-pair region of oxygen, and – to a lesser degree – in the non-bonding regions of two hydrogens. This effect on the heavy atom should indeed be expected for its promotion due to the nearly tetrahedral sp^3-hybridization.

[1.1.1]

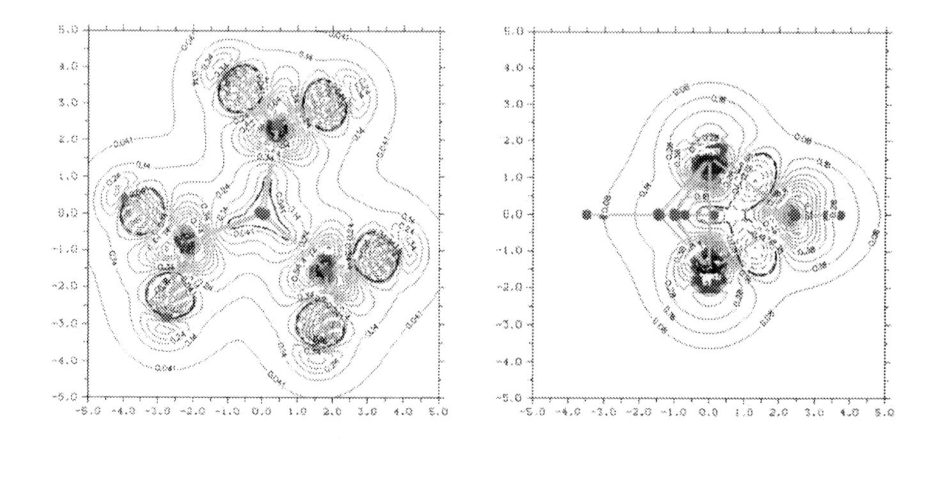

[2.1.1]

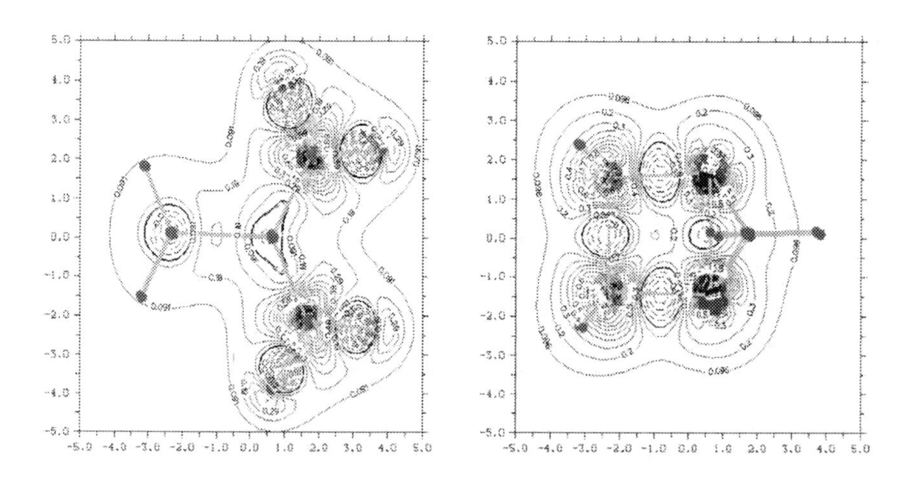

Figure 26. (beginning) The same as in Figure 17 for the four propellanes of Figure 5

[2.2.1]

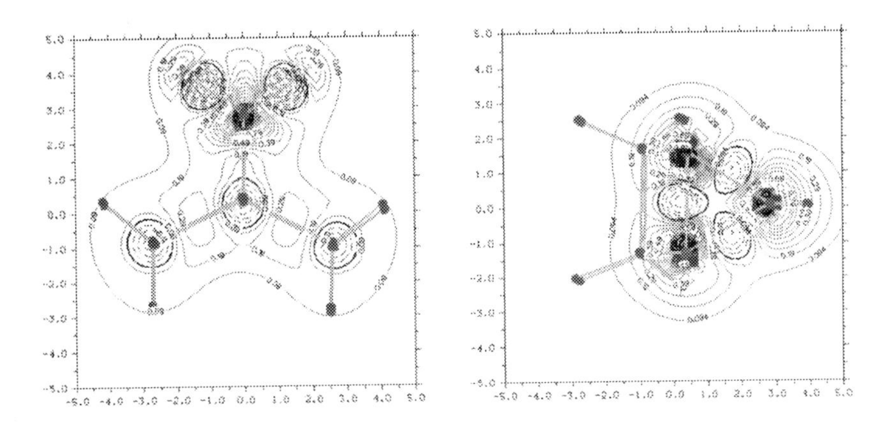

[2.2.2]

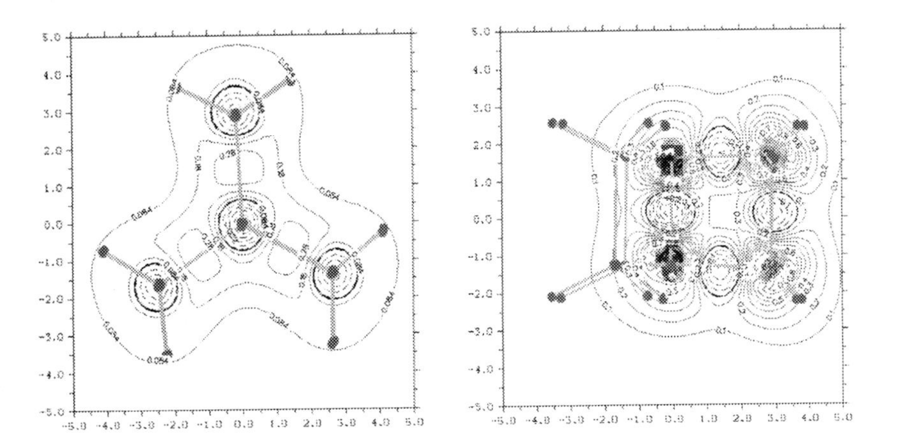

Figure 26. (conclusion)

The chemical bonds in small hydrocarbons are investigated in contour maps of Figures (22)-(25). These diagrams testify to the efficiency of the CG criterion in localizing all C—C and C—H bonding regions in ethane, ethylene, acetylene, butadiene, and benzene. The CG pattern of the triple bond between the carbon atoms in acetylene (upper panel of Figure 23) strongly resembles that observed in N_2 (Figure 20). In acetylene the two cylindrical bonding regions of the C—H bonds, axially extended due to a strong linear (*sp*-hybridization) promotion of both carbon atoms, and the central bonding basin due to the triple C—C bond, now transversely extended in the directions perpendicular to the bond axis, can be clearly seen in the contour map. The $f^{nadd.}(\boldsymbol{r}) > 0$ regions on each carbon atom, very much resembling the atomic $2p_\pi$ distributions, reflect the presence of the bond π component. The depletion of the $2p_\pi$ electron density near the carbon nuclei generates more structure in the electron π-donating (non-bonding) regions of both carbon atoms, and hence less structure (more delocalization) in the π-accepting (bonding) volume between the two nuclei.

The butadiene contour map in the molecular plane containing all nuclei is shown in the lower panel in Figure 23. As seen in this diagram all bonds are properly accounted for by the IT CG probe. The same conclusion follows from examining Figure 25, where the CG contour maps for benzene are shown, in the molecular plane (upper diagram) and in the perpendicular section containing the C—C bond (lower panel). It also follows from Figure 24, where the additional CG cuts for butadiene are shown, in the planes od sections perpendicular to the molecular plane, along the peripheral and middle C—C bonds, respectively, that the π bond between the neighboring peripheral carbons in butadiene is indeed stronger than its central counterpart, in full accord with familiar quantum chemical predictions from the SCF LCAO MO theory.

Finally, the bonding patterns in a series of four small propellanes of Figure 5 are examined in CG contour maps of Figure 26. Each row of the figure is devoted to a different propellane, arranged from the smallest [1.1.1] molecule, exhibiting three single-carbon bridges, to the largest [2.2.2] system, consisting of three double-carbon bridges; the left panel of each row corresponds to the plane of section perpendicular to the central bond between the bridgehead carbons, at the bond midpoint, while the axial cut of the right panel involves one of the system carbon bridges.

The main result of the previous studies, the apparent lack of the direct (through-space) bond between the carbon bridgeheads in the [1.1.1] and [2.1.1] systems, and a presence of practically single bond in the [2.2.1] and [2.2.2] propellanes, remain generally confirmed by the new CG probe, but this transition is now seen to be less sharp, with very small bonding basins between bridgeheads being also observed in the two smallest molecules. Thes, in accordance with the CG criterion the transition from the missing direct bonding in [1.1.1] system to the full central bond in [2.2.2] propellane appears to be less abrupt: a very small bonding basin identified in the former case is steadily evolving into that attributed to the full bond in both [2.2.1] and [2.2.2] propellanes. The C—C and C—H chemical bonds in the bridges are again perfectly delineated by the valence surfaces of the vanishing CG density.

The pattern of the non-additive Fisher information density always appears to be very much polarizational in character, with the closed bonding-regions of the negative CG being separated by the molecular environment of the positive values of this quantity, marking the system non-bonding regions.

5.5. Concluding Remarks

A depletion of the CG density in bonding regions is accompanied by increases of this non-additive information quantity in the nearby regions, so that there is but little net increase in the overall Fisher information content of the molecular system compared to that in the atomic promolecule. Obviously, the overall amount of the Fisher information in the molecule is not conserved [24] since different electron densities/probabilities give rise to different average Fisher information content related to the expectation value of the system kinetic energy of electrons. Since on average the molecule exhibits a higher degree of "structure", due to the dominating contraction of the free-atom electron distributions, it also exhibits a net increase in the overall Fisher information content, and hence also in the average kinetic energy of electrons, as indeed implied by the molecular virial theorem. More specifically, for the equilibrium geometry of the molecule, i.e., the vanishing forces acting on the system nuclei, the simple, atomic-like virial relation requires $\langle T\rangle = -\langle E\rangle$. Therefore, a small negative displacement in the average electronic energy of the molecule M relative to its atomic promolecule M^0 due to formation of the chemical bonds, $\Delta\langle E\rangle = \langle E\rangle - \langle E\rangle^0 < 0$, indeed implies the associated positive displacement in the system average kinetic energy, $\Delta\langle T\rangle = \langle T\rangle - \langle T\rangle^0 > 0$, which is proportional to the overall Fisher information: $\Delta\langle T\rangle \propto \Delta\langle I\rangle = \langle I\rangle - \langle I\rangle^0$. One also observes that in the minimum basis-set description of H_2 $\Delta\langle T\rangle \propto I^{c\text{-}g} > 0$ [25].

One of the primary goals of quantum chemistry is to identify the physical sources of the chemical bond. Most of existing theoretical interpretations of the origins of the covalent chemical bond at the molecular equilibrium geometry emphasize, almost exclusively, the potential (interaction) aspect of this phenomenon, focusing on the mutual attraction between the accumulation of electrons between the two atoms (the negative "*bond*-charge") and the partially screened (positively charged) nuclei. Indeed, the familiar virial-theorem decomposition of the diatomic BO potential indicates that for the equilibrium bond length it is the overall (negative) change in its potential component, due to a contraction of constituent atoms in the presence of each other, which is ultimately responsible for the net stabilizing (bonding) effect. The associated change in the overall kinetic energy is negative only at an earlier stage of the mutual approach by both atoms, when it is dominated by the *longitudinal* contribution associated with the gradient component along the bond axis. It ultimately assumes the destabilizing (antibonding) character at the equilibrium inter-nuclear separation, mainly due to its *transverse* contribution associated with the gradient components in the directions perpendicular to the bond axis. This overall virial theorem perspective also indicates that the kinetic energy constitutes the driving force of the bond-formation process. It follows from the classical analysis by Ruedenberg and co-workers [63,64] that the contraction of the atomic electron distributions is possible in molecule due to the relative lowering of the kinetic energy in the bonding region between the two nuclei, as reflected by both the longitudinal and transverse gradient components of the average electronic kinetic energy at large internuclear separations. Therefore, the whole process of redistributing electrons during formation of the chemical bond can be also regarded as being "catalyzed" by the gradient effect of the kinetic-energy. A similar conclusion follows from theoretical analysis by Goddard and Wilson [65].

The overall displacement in the kinetic energy (Fisher information) due to the chemical bond formation emphasizes the overall contraction of the electronic density in the presence of the remaining nuclear attractors in the molecule. It only blurs the picture of subtle information origins of chemical bonds, since the *total* kinetic energy component combines the delicate (truly bonding) *inter*-atomic effects originating from the stabilizing combinations of AO in the occupied MO, which ultimately determine the resultant bonding patterns in the molecule, and the accompanying processes of the *intra*-atomic polarization, which involve the non-bonding (lone) pairs of both the inner- and outer-shell electrons. Therefore, the displacement in the overall kinetic energy contribution effectively hides the minute changes in the system valence-shell, which are associated by chemists with the chemical bond concept. Some partitioning of this overall energy component is called for to separate the subtle *bonding* phenomena from the associated *non-bonding* promotion mechanism.

The ELF and CG criteria, which reflect the non-additive kinetic-energy/Fisher-information terms, adopt the complementary view by stressing the importance of the kinetic-energy bond component and allow for an identification of the bonding regions in the molecule. Only by focusing on the *non-additive* part of the electronic kinetic energy can one uncover the real information origins of chemical bonds and define the useful local probes for their localization and eventually – for the quantitative information descriptors based upon the kinetic-energy/Fisher-information [25b].

The CG concept, based upon the interference (non-additive) kinetic-energy terms, can be applied both locally, by using maps of the non-additive Fisher information in AO-resolution to identify and visualize the bonding regions in a molecule, and in the partially condensed forms obtained by integrating the local index over these basins of the physical space [25b]. Such numerical indices can be expected to provide sensitive tools for *quantitatively* describing the bonding patterns in molecules.

Since in the ELF concept the inverse of the key function $D_\sigma(\boldsymbol{r})$, proportional to the negative of non-additive Fisher information density in MO resolution, is a measure of the electron *localization*, $D_\sigma(\boldsymbol{r})$ itself reflects the electron *delocalization* effect. Therefore, the negative of $f_{MO}^{nadd.}(\boldsymbol{r})$ similarly provides the local index of the electron delocalization due to chemical bonds. The same applies to the AO resolution, in which the CG density is defined: the negative $f_{AO}^{nadd.}(\boldsymbol{r})$, which assumes positive values in bonding regions, measures a degree of electron delocalization in the system chemical bonds. Its reference, zero level indeed characterizes the non-bonding (triplet) configuration $\Psi_1 = [\varphi_b^1 \varphi_a^1]$ of Section 6.2:

$$
\begin{aligned}
{}^3\Psi_1(1,2) &= \frac{1}{\sqrt{2}}[\varphi_b(\boldsymbol{r}_1)\varphi_a(\boldsymbol{r}_2) - \varphi_b(\boldsymbol{r}_2)\varphi_a(\boldsymbol{r}_1)]\alpha(\sigma_1)\alpha(\sigma_2) \\
&= \frac{1}{\sqrt{2}}[A(\boldsymbol{r}_1)B(\boldsymbol{r}_2) - A(\boldsymbol{r}_2)B(\boldsymbol{r}_1)]\alpha(\sigma_1)\alpha(\sigma_2),
\end{aligned}
$$

in which the two sets of orbitals, $\boldsymbol{\chi} = (A, B)$ and $\boldsymbol{\varphi} = (\varphi_b, \varphi_a)$, are physically equivalent by defining the same (spatial) Slater determinant of the system wave-function.

Therefore, relative to this non-bonding configuration, the positive value of $D_{AO}(\boldsymbol{r}) = -f_{AO}^{nadd.}(\boldsymbol{r})$ in the bonding region of the ground state ${}^1\Psi_0$ implies an increase in the electron delocalization at these locations, while its negative value in this region for the anti-bonding configuration ${}^1\Psi_2$ signifies a lowering of the electron delocalization in the system, compared to the separated-atom limit. This observation fully accords with the familiar displacements in the MO densities and those exhibited by the densities of the stockholder atoms-in-molecules [7] in these electronic states. This conditional probability index also reflects the textbook rule of the proportionality between the electron delocalization, i.e., electron-“sharing” by bonded atoms, and the covalent character of the chemical bond.

The novel CG visualization tool should prove useful in exploring and describing the bonding patterns in controversial molecular systems, the bonding structure of which still remains a matter for scientific debate, with alternative bond criteria giving conflicting answers to the very qualitative question of the existence or non-existence of the disputed chemical bonds between the specified atoms in the molecular environment under consideration. We have illustrated such a use of this local probe in the representative problem of the central bond in propellanes, between the bridgehead carbon atoms.

Chapter 6

ORBITAL COMMUNICATION THEORY OF THE CHEMICAL BOND

6.1. INTRODUCTION

The key concept of CTCB is the molecular communication (information) channel (see Section 1.4), which can be constructed at alternative levels of resolving the electron probabilities into the underlying elementary "events" determining the channel inputs $\boldsymbol{a} = \{a_i\}$ and outputs $\boldsymbol{b} = \{b_j\}$. For example, they may involve the finding of an electron on the basis-set orbital (AO), AIM, molecular fragment, *etc.* Such communication channels can be generated within both the *local* and *condensed* descriptions of electronic probabilities. These networks describe the probability/information propagation in the molecule and can be characterized using standard entropic quantities developed in IT for real communication devices.

Due to the electron delocalization throughout the network of chemical bonds in a molecule the transmission of "signals" about the electron-assignment to the underlying elementary events of the resolution in question becomes randomly disturbed, thus exhibiting the communication "noise". Indeed, an electron initially attributed to the given atom/orbital in the channel "input" $\boldsymbol{a}$ (molecular, promolecular, or the "ensemble" tailored) can be later found with a non-zero probability at several locations in the molecular "output" $\boldsymbol{b}$. This feature of the electron delocalization is embodied in the conditional probabilities of the *outputs-given-inputs*, $\mathbf{P}(\boldsymbol{b}\,|\,\boldsymbol{a}) = \{P(b_j|a_i) \equiv P(j|i)\}$, which define the molecular information network.

Both *one*- and *two*-electron approaches have been devised to construct this matrix. The latter have used the joint probabilities of simultaneous events involving two electrons in a molecule, in the AIM input and output, respectively, to determine the network conditional probabilities, e.g., [7,28-31], while the former [8,20-22,32] constructs the relevant orbital probabilities using the bond-projected superposition-principle of quantum mechanics.

The *two*-electron CTCB treatment has been found to give rise to rather poor representation of the bond-differentiation patterns in molecules, which is decisively improved within the OCT approach using the AO-resolution. The latter scheme complements its earlier orbital implementation using the effective AO-promotion channel generated from the sequential cascade of the intermediate orbital-transformation stages [33]. Such consecutive cascades of elementary information systems have been used to represent the orbital-transformations and electron-excitations in the resultant propagations of the electron probabilities in molecules. The information-cascade approach also provides the probability-scattering perspective on atomic promotion due to the orbital hybridization [33b].

In OCT the conditional probabilities determining the molecular communication channel in the basis-function resolution follow from the quantum-mechanical superposition principle [45] supplemented by the "physical" projection onto the subspace of the system occupied MO which determine the molecular network of chemical bonds [20,32]. Both the molecule as a whole and its constituent subsystems can be adequately described using the OCT bond indices. The internal and external indices of molecular fragments (groups of AO) can be efficiently generated using the appropriate *reduction* of the molecular channel, by combining several outputs into a single molecular fragment.

In OCT formulation of CTCB the off-diagonal orbital communications have been shown to be proportional to the corresponding Wiberg [39] or related quadratic indices of the chemical bond [46-54]. The Wiberg-callibrated IT indices of diatomic interactions in molecules, generated using the *input*-weighted approach which adopts the flexible ("ensemble") input probabilities to probe the localized bond in the molecule, have been successfully implemented in the spin-*Restricted Hartree-Fock* (RHF) theory [8b]. The resulting IT descriptors been shown to account for the chemical intuition quite well, at the same time providing the resolution of the overall bond-multiplicities into the complementary IT-covalent and IT-ionic components. In the same study the need for recognizing the signs of the off-diagonal matrix elements of the CBO matrix has been stressed, in order to properly account for the so called "occupation" decoupling, when the *anti*-bonding MO become successively populated in the excited electron configurations. In this chapter we shall briefly summarize the theoretical basis and representative applications of OCT to the localized chemical bonds and the bond coupling-phenomena in molecular and/or reactive systems.

6.2. MOLECULAR INFORMATION SYSTEM IN ATOMIC ORBITAL RESOLUTION

In MO theory the network of chemical bonds is determined by the occupied MO in the system ground-state. Let us assume, for simplicity, the closed-shell (*cs*) electron configuration of $N = 2n$ electronic system, within the standard RHF description, which involves the n lowest (doubly occupied, orthonormal) MO. In the familiar SCF LCAO MO approach they are generated as linear combinations of the adopted AO (basis functions) $\boldsymbol{\chi} = (\chi_1, \chi_2, \ldots, \chi_m) = \{\chi_i\}$, $\langle \boldsymbol{\chi} | \boldsymbol{\chi} \rangle = \{\delta_{i,j}\} \equiv \mathbf{I}$, e.g., Löwdin's symmetrically-orthogonalized AO (OAO): $\boldsymbol{\varphi} = (\varphi_1, \varphi_2, \ldots, \varphi_n) = \{\varphi_s\} = \boldsymbol{\chi}\mathbf{C}$; here the rectangular matrix $\mathbf{C} = \{C_{i,s}\} = \langle \boldsymbol{\chi} | \boldsymbol{\varphi} \rangle$ groups the relevant expansion coefficients to be determined using the iterative SCF procedure.

The system electron density $\rho(\boldsymbol{r})$, and hence also the *one*-electron probability distribution $p(\boldsymbol{r}) = \rho(\boldsymbol{r})/N$, the *shape*-factor of ρ, are determined by the CBO matrix $\boldsymbol{\gamma}$ of Eqs (187) and (196),

$$\boldsymbol{\gamma} = 2\langle \boldsymbol{\chi} | \boldsymbol{\varphi} \rangle \langle \boldsymbol{\varphi} | \boldsymbol{\chi} \rangle = 2\mathbf{C}\mathbf{C}^{\dagger} \equiv 2\langle \boldsymbol{\chi} | \hat{\mathrm{P}}_{\varphi} | \boldsymbol{\chi} \rangle = \{\gamma_{i,j} = 2\langle \chi_i | \hat{\mathrm{P}}_{\varphi} | \chi_j \rangle \equiv 2\langle i | \hat{\mathrm{P}}_{\varphi} | j \rangle\}, \tag{197}$$

the AO representation of the projection operator $\hat{\mathrm{P}}_{\varphi} = |\boldsymbol{\varphi}\rangle\langle\boldsymbol{\varphi}| = \sum_s |\varphi_s\rangle\langle\varphi_s| \equiv \sum_s \hat{\mathrm{P}}_s$ onto the subspace of all doubly-occupied MO:

$$\rho(\boldsymbol{r}) = 2\boldsymbol{\varphi}(\boldsymbol{r})\boldsymbol{\varphi}^{\dagger}(\boldsymbol{r}) = \boldsymbol{\chi}(\boldsymbol{r})\,[2\mathbf{C}\mathbf{C}^{\dagger}]\,\boldsymbol{\chi}^{\dagger}(\boldsymbol{r}) \equiv \boldsymbol{\chi}(\boldsymbol{r})\,\boldsymbol{\gamma}\,\boldsymbol{\chi}^{\dagger}(\boldsymbol{r}) = Np(\boldsymbol{r}), \tag{198}$$

Therefore, this density matrix satisfies the familiar idempotency relation:

$$(\boldsymbol{\gamma})^2 = 4\langle \boldsymbol{\chi} | \hat{\mathrm{P}}_{\varphi} | \boldsymbol{\chi} \rangle \langle \boldsymbol{\chi} | \hat{\mathrm{P}}_{\varphi} | \boldsymbol{\chi} \rangle = 4\langle \boldsymbol{\chi} | \hat{\mathrm{P}}_{\varphi}^2 | \boldsymbol{\chi} \rangle = 4\langle \boldsymbol{\chi} | \hat{\mathrm{P}}_{\varphi} | \boldsymbol{\chi} \rangle = 2\boldsymbol{\gamma}. \tag{199}$$

The CBO (density) matrix reflects the promoted, *valence*-state of AO in the molecule, with the diagonal elements measuring the effective electron occupations, $\{N_i = \gamma_{i,i} = Np_i\}$, and hence also net charges of the basis functions, with probabilities $\boldsymbol{p} = \{p_i = \gamma_{i,i}/N\}$ of the AO being occupied in the molecule, $\sum_i p_i = 1$. The signs of its off-diagonal (*inter*-atomic) elements, called bond-orders, reflect the character of the effective chemical interaction between the given pair of AO, positive - for the bonding interaction, and negative - for a resultant anti-bonding AO coupling in all MO, with the zero value identifying the mutually non-bonding status of the two basis functions.

The molecular information channel in the (condensed) orbital resolution involves the AO events χ in its input $\boldsymbol{a} = \{\chi_i\}$ and output $\boldsymbol{b} = \{\chi_j\}$. It represents the effective communication promotion of these basis functions in the molecule *via* the probability/information scattering described by the conditional probabilities of the AO-outputs given the AO-inputs, identified by the *row* (input) and *column* (output) indices, respectively. In this *one*-electron description the AO→AO communication network is thus determined by the conditional probabilities of the output AO-events, given the input AO-events,

$$\mathbf{P}(\boldsymbol{b}|\boldsymbol{a}) = \{P(\chi_j|\chi_i) \equiv P(j|i) = P(i\wedge j)/p_i\}, \qquad \sum_j P(j|i) = 1, \tag{200}$$

where the associated joint probabilities of simultaneously observing two AO in the system chemical bonds $\mathbf{P}(\boldsymbol{a}\wedge\boldsymbol{b}) = \{P(i\wedge j)\}$ satisfy the usual partial and total normalization relations:

$$\sum_i P(i\wedge j) = p_j, \qquad \sum_j P(i\wedge j) = p_i, \qquad \sum_i\sum_j P(i\wedge j) = 1. \tag{201}$$

The conditional probabilities $\mathbf{P}(\boldsymbol{b}|\boldsymbol{a})$ define the information scattering in the AO-promotion channel of the molecule, in which the "signals" of the molecular electron allocations to basis functions are transmitted between the AO inputs and outputs. Such communication system constitutes the basis of OCT of the chemical bond.

As argued elsewhere [20,32], by using the generalized, bond-projected *superposition principle* of quantum mechanics (see Appendix B), this matrix of the (*physical*) conditional probabilities involves the squares of corresponding elements of the CBO matrix:

$$\mathbf{P}(\boldsymbol{b}|\boldsymbol{a}) = \{P(j|i) = \mathcal{N}_i \left|\langle i|\hat{\mathrm{P}}_\varphi|j\rangle\right|^2 = (2\gamma_{i,i})^{-1}\gamma_{i,j}\gamma_{j,i}\}, \tag{202}$$

where the closed-shell normalization constant $\mathcal{N}_i = (2\gamma_{i,i})^{-1}$ follows directly from Eq. (199) (for the open-shell generalization see [8b]). These probabilities explore the dependencies between AO resulting from their simultaneous participation in the framework of the occupied MO, i.e. their involvement in the entire network of chemical bonds in the molecule. This molecular channel can be probed using both the promolecular ($\boldsymbol{p}^0 = \{p_i^0\}$), molecular ($\boldsymbol{p}$), or arbitrary, e.g., ensemble input probabilities, in order to extract the desired IT descriptors of the fragment or global bond multiplicities and their *ionic* and *covalent* components [7].

In this approach the off-diagonal conditional probability of *j*th AO-output given *i*th AO-input is thus proportional to the squared element of the CBO matrix linking the two AO, $\gamma_{j,i} = \gamma_{i,j}$, thus being also proportional to the corresponding AO contribution $\mathcal{M}_{i,j} = \gamma_{i,j}^2$ to the Wiberg [39] index of the overall chemical bond-order between two atoms A and B in the molecule [53],

$$\mathcal{M}(\text{A, B}) = \Sigma_{i \in \text{A}} \Sigma_{j \in \text{B}} \mathcal{M}_{i,j}, \tag{203}$$

or to generalized quadratic descriptors of molecular bond multiplicities [46-61].

It can be straightforwardly verified using the idempotency relation of Eq. (199) that the associated joint probability matrix,

$$\mathbf{P}(\boldsymbol{a}\wedge\boldsymbol{b}) = \{P(i\wedge j) = p_i\, P(j|i) = (2N)^{-1}\gamma_{i,j}\gamma_{j,i} = (2N)^{-1}\langle i|\hat{\mathrm{P}}_\varphi|j\rangle\langle j|\hat{\mathrm{P}}_\varphi|i\rangle\}, \tag{204}$$

indeed satisfies the normalization conditions of Eq. (201), e.g.,

$$\Sigma_i\, P(i\wedge j) = (2N)^{-1}\Sigma_i\, \gamma_{j,i}\gamma_{i,j} = (2N)^{-1}\, 2\gamma_{j,j} = p_j. \tag{205}$$

6.3. Covalent and Ionic Descriptors of Chemical Bonds

In OCT the entropy/information indices of the covalent/ionic components of all chemical bonds in the given molecular system as a whole represent the complementary descriptors of the average communication *noise* and the amount of information *flow* in the molecular information channel. The molecular input signal $\boldsymbol{P}(\boldsymbol{a}) \equiv \boldsymbol{p}$ generates the same distribution in the output of the molecular channel,

$$\boldsymbol{p}\,\mathbf{P}(\boldsymbol{b}|\boldsymbol{a}) = \{\Sigma_i p_i\, P(j|i) \equiv \Sigma_i P(i\wedge j) = p_j\} = \boldsymbol{p}, \tag{206}$$

thus identifying $\boldsymbol{p}$ as the *stationary* probability vector of AO in the molecular ground state, while the promolecular input $\boldsymbol{P}(\boldsymbol{a}^0) \equiv \boldsymbol{p}^0$ in general produces different output probability. The purely molecular communication channel, with $\boldsymbol{p}$ defining its input signal, is devoid of any reference (history) of the chemical bond formation and generates the average-noise index of the molecular IT bond-*covalency*, measured by the *conditional-entropy* of the system outputs given inputs (see Sections 1.3 and 1.4):

$$S(\boldsymbol{P}(\boldsymbol{b})|\boldsymbol{P}(\boldsymbol{a})) \equiv H(\mathbf{B}|\mathbf{A}) = -\sum_i \sum_j P(i\wedge j)\log[P(i\wedge j)/p_i]$$

$$= \sum_i p_i[-\sum_j P(j|i) \log P(j|i)] \equiv \sum_i p_i S_i \equiv S(\boldsymbol{p}|\boldsymbol{p}) \equiv S. \tag{207}$$

This average-noise descriptor expresses the difference between the Shannon entropies of the molecular *one*- and *two*-orbital probabilities [see Eq. (17) and Schemes 1, 4],

$$S = H(\mathbf{AB}) - H(\mathbf{A}),$$

$$H(\mathbf{A}) = -\sum_i p_i \log p_i \equiv S(\boldsymbol{P}(\boldsymbol{a})) = H(\mathbf{B}) \equiv S(\boldsymbol{P}(\boldsymbol{b})) ,$$

$$H(\mathbf{AB}) \equiv S(\mathbf{P}(\boldsymbol{a}\wedge\boldsymbol{b})) = -\sum_i \sum_j P(i\wedge j) \log P(i\wedge j). \tag{208}$$

Hence, for the *independent* input and output events, when $\mathbf{P}^{ind.}(\boldsymbol{a}\wedge\boldsymbol{b}) = \{p_i p_j\}$, $S(\mathbf{P}^{ind.}(\boldsymbol{a}\wedge\boldsymbol{b})) = 2S(\boldsymbol{p})$, and hence $S^{ind.} = S(\boldsymbol{p})$.

The AO channel with $\boldsymbol{P}(\boldsymbol{a}^0) = \boldsymbol{p}^0$ determining its input "signal" probability refers to the initial state in the bond-formation process - the system *promolecule*. It corresponds to the ground-state (fractional) occupations of the AO contributed by the system constituent free atoms, before their mixing into MO. This input signal gives rise to the average information-flow descriptor of the system IT bond-*ionicity*, given by the *mutual-information* in the channel inputs and outputs:

$$I(\boldsymbol{P}(\boldsymbol{a}^0){:}\boldsymbol{P}(\boldsymbol{b})) \equiv I(\mathbf{A}^0{:}\mathbf{B}) = \sum_i\sum_j P(i\wedge j)\log[P(i\wedge j)/(p_j p_i^0)]$$

$$= \sum_i p_i\{\sum_j P(j|i) \log[P(i|j)/p_i^0]\} \equiv \sum_i p_i I_i \equiv I(\boldsymbol{p}^0{:}\boldsymbol{p}) \equiv I$$

$$= S(\boldsymbol{P}(\boldsymbol{b})) + S(\boldsymbol{P}(\boldsymbol{a}^0)) - S(\mathbf{P}(\boldsymbol{a}\wedge\boldsymbol{b})) = S(\boldsymbol{p}^0) - S. \tag{209}$$

This *amount of information* reflects the fraction of the initial (promolecular) information content $S(\boldsymbol{p}^0)$, which has not been dissipated as noise in the molecular communication system and is received in the channel output. In particular, for the molecular input, when $\boldsymbol{p}^0 = \boldsymbol{p}$,

$$I(\boldsymbol{P}(\boldsymbol{a}) : \boldsymbol{P}(\boldsymbol{b})) = \sum_i \sum_j P(i,j) \log[P(i,j)/(p_j p_i)] = S(\boldsymbol{p}) - S \equiv I(\boldsymbol{p} : \boldsymbol{p}). \quad (210)$$

Hence, for the independent input and output events $I^{ind.}(\boldsymbol{P}(\boldsymbol{a}) : \boldsymbol{P}(\boldsymbol{b})) = 0$.

Finally, the sum of these two bond components,

$$\mathcal{N}(\boldsymbol{P}(\boldsymbol{a}^0); \boldsymbol{P}(\boldsymbol{b})) = S + I \equiv \mathcal{N}(\boldsymbol{p}^0; \boldsymbol{p})$$

$$\equiv \mathcal{N} = S(\boldsymbol{p}^0) = \sum_i p_i (S_i + I_i) \equiv \sum_i p_i \mathcal{N}_i, \quad (211)$$

where $\mathcal{N}_i = -\log p_i^0$ stands for the self-information in the promolecular AO-intput event χ_i, measures the overall IT bond-multiplicity of all bonds in the molecular system under consideration. It is seen to be conserved at the promolecular-entropy level, which marks the initial information content of AO probabilities. Alternatively, for the molecular input, when $\boldsymbol{P}(\boldsymbol{a}) = \boldsymbol{p}$, this quantity preserves the Shannon entropy of the molecular input probabilities:

$$\mathcal{N}(\boldsymbol{P}(\boldsymbol{a}); \boldsymbol{P}(\boldsymbol{b})) = S(\boldsymbol{P}(\boldsymbol{b})|\boldsymbol{P}(\boldsymbol{a})) + I(\boldsymbol{P}(\boldsymbol{a}) : \boldsymbol{P}(\boldsymbol{b})) = S(\boldsymbol{P}(\boldsymbol{a})) = S(\boldsymbol{p}). \quad (212)$$

We recall (see Schemes 1 and 4) that for two dependent probability schemes, the common (overlap) area of the associated entropy circles corresponds to the mutual information in both distributions, $I(\boldsymbol{P}(\boldsymbol{a}):\boldsymbol{P}(\boldsymbol{b}))$, while the remaining parts of individual circles represent the corresponding conditional entropies $S(\boldsymbol{P}(\boldsymbol{b})|\boldsymbol{P}(\boldsymbol{a}))$ or $S(\boldsymbol{P}(\boldsymbol{a})|\boldsymbol{P}(\boldsymbol{b}))$. The latter measure the residual uncertainty about events in one set, when one has the full knowledge of the occurrence of the events in the other set of events. Accordingly, the area enclosed by the envelope of these two overlapping circles represents the entropy in the joint distribution of these two sets of outcomes:

$$S(\boldsymbol{P}(\boldsymbol{a}) \wedge \boldsymbol{P}(\boldsymbol{b})) = S(\boldsymbol{P}(\boldsymbol{a})) + S(\boldsymbol{P}(\boldsymbol{b})) - I(\boldsymbol{P}(\boldsymbol{a}):\boldsymbol{P}(\boldsymbol{b}))$$

$$= S(\boldsymbol{P}(\boldsymbol{a})) + S(\boldsymbol{P}(\boldsymbol{b})|\boldsymbol{P}(\boldsymbol{a}))$$

$$= S(\boldsymbol{P}(\boldsymbol{b})) + S(\boldsymbol{P}(\boldsymbol{a})|\boldsymbol{P}(\boldsymbol{b})). \quad (213)$$

6.4. TWO-ORBITAL MODEL OF CHEMICAL BOND

To illustrate these IT concepts let us again examine the 2-AO model of the chemical bond (see Section 5.3). The ground-state density matrix γ_0 [Eq. (188)], for the doubly occupied bonding MO φ_b, generates the following conditional probability matrix $\mathbf{P}(\boldsymbol{b}|\boldsymbol{a}) = \mathbf{P}(\chi|\chi) = \{P(j|i)\}$ [Eq. (202)]:

$$\mathbf{P}(\chi|\chi) = \begin{bmatrix} P & Q \\ P & Q \end{bmatrix}, \tag{214}$$

which determines the AO→AO communication network, shown in Scheme 6, for this model bond system. In this channel one then adopts the molecular input signal, $\boldsymbol{p} = (P, Q = 1 - P)$, to extract the bond IT-covalency index, which measures the channel average communication noise, and the promolecular input signal, $\boldsymbol{p}^0 = (\frac{1}{2}, \frac{1}{2})$, to calculate the IT-ionicity index measuring the channel information capacity relative to this *covalent* promolecule, in which basis functions contribute a single electron each to form the chemical bond.

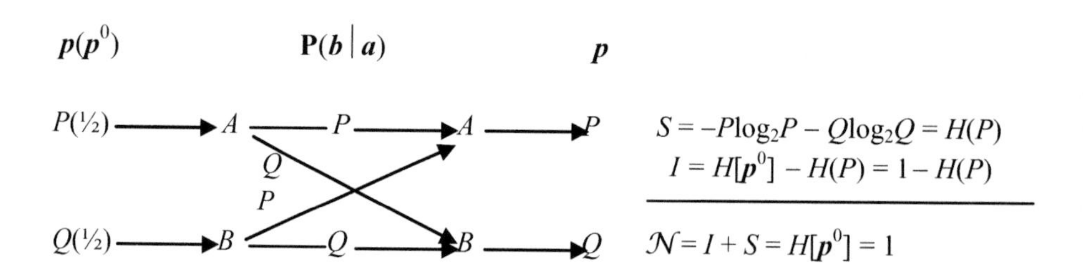

Scheme 6. Communication channel of the 2-OAO model of the chemical bond and its entropy/information descriptors (in bits)

The bond IT covalency $S(P)$ is then determined by the binary entropy function $H(P) = -P\log_2 P - Q\log_2 Q = H[\boldsymbol{p}]$ of Figure 1. It reaches the maximum value $H(P = \frac{1}{2}) = 1$bit for the symmetric bond $P = Q = \frac{1}{2}$, e.g., the σ bond in H_2 or the π bond in ethylene, and vanishes for the *lone*-pair molecular configurations, when $P = (0, 1)$, $H(P = 0) = H(P = 1) = 0$, marking the alternative *ion*-pair configurations A^+B^- and A^-B^+, respectively, relative to the initial AO occupations $\boldsymbol{N}^0 = (1, 1)$ in the assumed promolecular reference.

The complementary descriptor of the IT-ionicity, determining the channel mutual information (*capacity*), $I(P) = H[\mathbf{p}^0] - H(P) = 1 - H(P)$, reaches the highest value for these two limiting electron-transfer configurations $P = (0, 1)$: $I(P = 0) = I(P = 1) = H(½) = 1$ bit. This ionicity descriptor is thus seen to identically vanish for the purely-covalent, symmetric bond, $I(P = ½) = 0$.

Therefore, these two bond-multiplicity components yield the conserved overall IT bond index $\mathcal{N}(P) = S(P) + I(P) = 1$ bit, marking a single bond in IT, in the whole range of admissible bond polarizations $P \in [0, 1]$. Thus, this simple model properly accounts for the competition between the bond covalency and ionicity, while preserving the single bond-order measure reflected by the conserved overall IT multiplicity of the chemical bond. Similar effects transpire from the *two*-electron CTCB [7] and the quadratic bond indices formulated in the MO theory [46-52].

6.5. Bond-Weighted Channels and Wiberg Index

In typical SCF LCAO MO calculations the lone pairs of the *valence*- and/or *inner*-shell electrons can strongly affect the IT descriptors of chemical bonds. It has been argued elsewhere [8,21] that the elimination of such *lone*-pair contributions to the resultant IT bond indices of diatomic fragments in molecules requires an *ensemble* approach, in which the input probabilities are derived from the *joint* bond-probabilities of two AO centered on different atoms. Indeed, the contributions due to each AO input should be weighted using the corresponding joint (*two*-orbital) probabilities, which reflect the actual simultaneous participation of the given pair of basis functions in the system chemical bonds, thus effectively projecting out the spurious contributions due to the inner- and outer-shell AO, which are excluded from mixing into the delocalized MO combinations. This probability-weighting procedure, known as the flexible input approach, has been shown to be capable of reproducing the Wiberg bond order in diatomics [8b,21], at the same time providing the IT-covalent/ionic resolution of this index.

The localized (diatomic) bond multiplicities in molecules are mainly determined by the constituent AO of both atoms, $\chi_{AB} = (\chi_A, \chi_B)$. This partial basis corresponds to the diatomic block $\gamma_{AB} = \{\gamma_{X,Y}, (X, Y) = A, B\}$ of the molecular density matrix and the associated part in the molecular matrix of the conditional probabilities between AO contributed by both atoms: $\mathbf{P}_{AB}(\chi_{AB}|\chi_{AB}) = \{\mathbf{P}(\chi_Y|\chi_X),$

(X,Y) = A, B}. The former determines the effective number of electrons on AB in the molecule given by the partial trace $N_{AB} = \sum_{i \in AB} \gamma_{i,i}$.

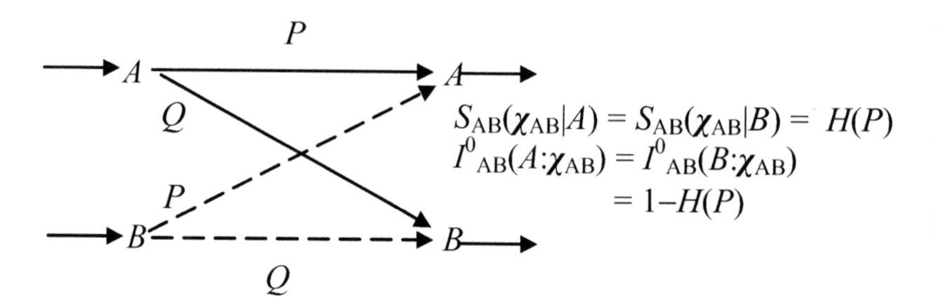

Scheme 7. The elementary (*row*) *sub*-channels [7] due to inputs *A* (solid lines) and *B* (broken lines) in the 2-OAO model of the chemical bond

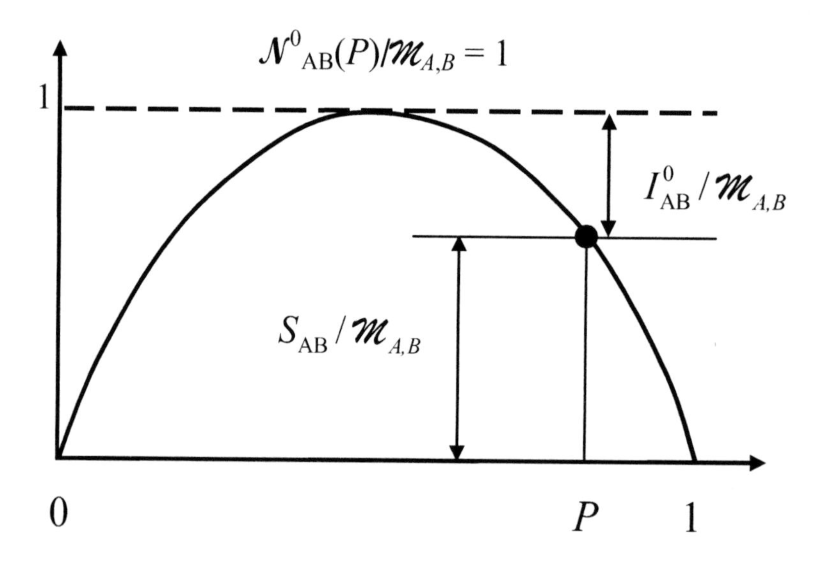

Figure 27. Variations of the IT-covalent $[S_{AB}(P)]$ and and IT-ionic $[I^0_{AB}(P)]$ components (in $\mathcal{M}_{A,B}$ units) of the chemical bond in the 2-OAO model (see Schemes 6 and 7) with changing MO polarization P and the conservation of the relative total bond-order $\mathcal{N}^0_{AB}(P)/\mathcal{M}_{A,B} = 1$.

We begin this section by applying this weighting procedure to the 2-OAO model of the preceding section. In the bond-weighted approach one distinguishes in the molecular channel of Scheme 6 the elementary (*row*) *sub*-channels due to each AO input (see Scheme 7). The conditional-entropy and mutual-information quantities for these partial communication systems, $\{S_{AB}(\chi_{AB}|i)\}, I^0_{AB}(i{:}\chi_{AB}); i = A, B\}$, respectively, with the latter being determined for the *covalent*-reference probabilities $\boldsymbol{p}^0 = (½, ½)$ marking the single electrons contributed by each AO to the diatomic chemical bond, are also listed in the diagram. Since the row-descriptors represent the IT indices *per electron* these contributions have to be multiplied by $N_{AB} = 2$ in the corresponding resultant measures.

Therefore, using the *off*-diagonal joint probability $P(A\wedge B) = P(B\wedge A) = PQ = \gamma_{A,B}\gamma_{B,A}/4$ as the ensemble probability for both OAO inputs gives the following average quantities for the model diatomic bond (see Figure 1):

$$S_{AB} = N_{AB}[P(A\wedge B)\, S_{AB}(\chi_{AB}|A) + P(B\wedge A)\, S_{AB}(\chi_{AB}|B)]$$

$$= 4PQ\, H(P) = \mathcal{M}_{A,B}\, H(P),$$

$$I^0_{AB} = N_{AB}[P(A\wedge B)\, I^0_{AB}(A{:}\chi_{AB}) + P(B\wedge A)\, I^0_{AB}(B{:}\chi_{AB})]$$

$$= 4PQ[1 - H(P)] = \mathcal{M}_{A,B}[1 - H(P)],$$

$$\mathcal{N}^0_{AB} = S_{AB} + I^0_{AB} = 4PQ = (\gamma_{A,B})^2 = \mathcal{M}_{A,B}. \tag{215}$$

We have thus recovered the Wiberg index as the overall IT descriptor of the chemical bond in 2-OAO model, $\mathcal{N}^0_{AB} = \mathcal{M}_{A,B}$, at the same time establishing its covalent, $S_{AB} = \mathcal{M}_{A,B}H(P)$, and ionic, $I^0_{AB} = \mathcal{M}_{A,B}[1 - H(P)]$, contributions. It follows from Figure 27 that these IT-covalency and IT-ionicity components compete with one another while conserving the Wiberg bond-order of the model as the overall information measure (in bits) of the model bond multiplicity.

This development can be straightforwardly generalized to a general case of several basis functions contributed by each bonded atom [8b,21]. The molecular probability scattering in the specified diatomic fragment (A, B) involving the basis functions $\chi_{AB} = (\chi_A, \chi_B)$ contributed by these two atoms to the overall set of AO, $\chi = \{\chi_X\}$, is fully characterized by the corresponding block $\mathbf{P}_{AB}(\chi_{AB}|\chi_{AB})$ of the molecular conditional probability matrix. It contains only the *intra*-diatomic communications, missing the probability propagations originating from AO of the remaining constituent atoms $\chi_Z \notin \chi_{AB}$, thus being perfectly capable of describing the *localized* chemical interactions between A and B.

The atomic *output*-reduction [7] of $\mathbf{P}(\chi_{AB}|\chi_{AB})$, carried out by combining the AO events χ_X into a single atomic event X in the output of the molecular channel, gives the associated *condensed* conditional probabilities of such partially-reduced information system of the diatomic fragment in question:

$$\boldsymbol{P}_{AB}(\boldsymbol{X}_{AB}|\chi_{AB}) = [\boldsymbol{P}(A|\chi_{AB}), \boldsymbol{P}(B|\chi_{AB})] = \{P(X|i)\} = \sum_{j\in X} \boldsymbol{P}(j|\chi_{AB})\,;$$

$$\chi_i \in \chi_{AB}, \quad \boldsymbol{X}_{AB} = \{X = A, B\}. \qquad (216)$$

Here, $P(X|i)$ measures the conditional probability that an electron originating from χ_i will be found on atom X in the molecule. The sum of these conditional probabilities over all AO contributed by the two atoms then determines the communication connections $\{P(AB|i)\}$ linking the condensed diatomic output AB and the given AO input χ_i in the communication system of the diatomic fragment under consideration:

$$\boldsymbol{P}(A|\chi_{AB}) + \boldsymbol{P}(B|\chi_{AB}) = \boldsymbol{P}(AB|\chi_{AB})$$

$$= \{P(AB|i) = P(A|i) + P(B|i) = \sum_{j\in(A,B)} P(j|i) \le 1\}. \qquad (217)$$

In other words, $P(AB|i)$ measures the probability that an electron occupying χ_i will be detected in the diatomic fragment AB of the molecule. The inequality in the preceding equation reflects the fact that the atomic basis functions participate in chemical bonds with *all* constituent atoms, with the equality sign thus corresponding only to the diatomic molecule, when $\chi_{AB} = \chi$.

The associated *fragment*-normalized AO probabilities,

$$\tilde{\boldsymbol{p}}(AB) = \{\tilde{p}_i(AB) = \gamma_{i,i}/N_{AB}, \quad \chi_i \in \chi_{AB}\}, \qquad \sum_{i\in(A,B)} \tilde{p}_i(AB) = 1, \qquad (218)$$

where $N_{AB} = \sum_{i\in(A,B)} \gamma_{i,i}$ stands for the number of electrons found in the molecule on this diatomic fragment and $\tilde{p}_i(AB)$ denotes the probability that one of them occupies $\chi_{i\in(A,B)}$, then determine the simultaneous probabilities of the joint two-orbital events:

$$\mathbf{P}_{AB}(\chi_{AB}\wedge\chi_{AB}) = \{P_{AB}(i\wedge j) = \tilde{p}_i(AB)\,P(j|i) = \gamma_{i,j}\gamma_{j,i}/(2N_{AB})\}. \qquad (219)$$

They in turn generate, *via* the relevant partial summations, the joint atom-orbital probabilities in AB, $\{P_{AB}(X, i)\}$:

$$\boldsymbol{P}_{AB}(\boldsymbol{X}_{AB}\wedge\boldsymbol{\chi}_{AB}) = [\boldsymbol{P}_{AB}(A\wedge\boldsymbol{\chi}_{AB}), \boldsymbol{P}_{AB}(B\wedge\boldsymbol{\chi}_{AB})]$$

$$= \{P_{AB}(X\wedge i) = \sum_{j\in X} P_{AB}(i\wedge j) \equiv \tilde{p}_i(AB)\, P(X|i),\quad X = A, B\}. \tag{220}$$

For the closed-shell molecular system one thus finds:

$$\boldsymbol{P}_{AB}(X\wedge\boldsymbol{\chi}_{AB}) = \left\{ P_{AB}(X\wedge i) = \tilde{p}_i(AB)\sum_{j\in X} P(j|i) = \sum_{j\in X}\frac{\gamma_{i,j}\gamma_{j,i}}{2N_{AB}} \right\} \equiv \boldsymbol{P}_{AB}(\boldsymbol{\chi}_{AB}\wedge X)^{T},$$

$$X = A, B. \tag{221}$$

These vectors of AO probabilities in the diatomic fragment AB subsequently define the condensed probabilities $\{P_X(AB)\}$ of both bonded atoms in this diatomic subsystem:

$$P_X(AB) = \frac{N_X(AB)}{N_{AB}} = \sum_{i\in(A,B)} P_{AB}(X\wedge i) = \sum_{i\in(A,B)}\sum_{j\in X}\frac{\gamma_{i,j}\gamma_{j,i}}{2N_{AB}},\quad X = A, B, \tag{222}$$

where the effective number of electrons $N_X(AB)$ on atom $X \in (A, B)$ now reads:

$$N_X(AB) = \sum_{i\in(A,B)}\sum_{j\in X}\frac{\gamma_{i,j}\gamma_{j,i}}{2}. \tag{223}$$

Therefore, in diatomic molecules, for which $\chi_{AB} = \chi$, one finds using the idempotency relation of Eq. (199),

$$P_X(AB) = \sum_{j\in X}\left(\sum_i \frac{\gamma_{j,i}\gamma_{i,j}}{2N_{AB}}\right) = \sum_{j\in X}\frac{\gamma_{j,j}}{N_{AB}} = \sum_{j\in X}\tilde{p}_j(AB),\quad X = A, B, \tag{224}$$

and hence $P_A(AB) + P_B(AB) = 1$. The relative importance of the basis functions contributed by one atom in forming the chemical bonds with the other atom of the specified diatomic fragment is reflected by the joint *bond* (b) probabilities of the two atoms, defined only by the diatomic components of the simultaneous probabilities [8b,21]:

$$P_b(\mathrm{A}\wedge\mathrm{B}) \equiv \sum_{j\in\mathrm{B}} P_{\mathrm{AB}}(\mathrm{A}\wedge j) \equiv \sum_{i\in\mathrm{A}} P_{\mathrm{AB}}(i\wedge\mathrm{B}) = P_b(\mathrm{B}\wedge\mathrm{A}) = \sum_{i\in\mathrm{A}}\sum_{j\in\mathrm{B}} \frac{\gamma_{i,j}\gamma_{j,i}}{2N_{\mathrm{AB}}}. \quad (225)$$

Indeed, the joint atom-orbital bond probabilities, $\{P_{\mathrm{AB}}(\mathrm{A}\wedge j),\ j\in\mathrm{B}\}$ and $\{P_{\mathrm{AB}}(i\wedge\mathrm{B},),\ i\in\mathrm{A}\}$, to be used as weighting factors in determining the average conditional-entropy (covalency) and mutual-information (ionicity) descriptors of the chemical bond(s) between A and B, assume appreciable magnitudes only when the electron occupying the atomic orbital χ_i of one atom is simultaneously found with a significant probability on the other atom, thus effectively excluding the contributions to the entropy/information bond descriptors due to the lone-pair electrons.

The *reference* bond probabilities of AO, to be used to calculate the mutual-information (IT-ionicity) bond index of the diatomic channel, have to be normalized to the corresponding sums $\boldsymbol{P}(\mathrm{AB}|\chi_{\mathrm{AB}}) = \{P(\mathrm{AB}|i)\}$ of Eq. (217). Since the bond probability concept of Eq. (225) involves symmetrically the two bonded atoms, one applies the same symmetry requirement in determining the associated reference bond probabilities of AO:

$$\{p_b(i) = P(\mathrm{AB}|i)/2;\quad i\in(\mathrm{A},\mathrm{B})\}, \quad (226)$$

where $P(\mathrm{AB}|i)$ denotes the probability that an electron originating from orbital χ_i will be found on atom A or B in the molecule.

As we have already indicated in Section 6.3, in OCT the complementary quantities characterizing the average *noise* (conditional entropy of the channel output given input) and the information *flow* (mutual information in the channel output and the reference input) in the diatomic communication system defined by the AO conditional probabilities provide the overall descriptors of the fragment bond covalency and ionicity, respectively. Both molecular and promolecular reference (input) probability distributions have been used in the past to determine the information index characterizing the displacement (ionicity) aspect of the system chemical bonds. In the bond-weighted diatomic development the equal bond probabilities of Eq.(226) will be used as the input reference values for this purpose.

In the A—B fragment development we similarly define the following ("ensemble") average contributions of both constituent atoms to the diatomic-covalency (delocalization) entropy:

$$S_{AB}(B|\chi_A) = \sum_{i \in A} P_{AB}(i \wedge B)\ S_{AB}(\chi_{AB}|i),$$
$$S_{AB}(A|\chi_B) = \sum_{i \in B} P_{AB}(i \wedge A)\ S_{AB}(\chi_{AB}|i), \tag{227}$$

where the Shannon entropy (in bits) of the conditional probabilities for the given AO input $\chi_i \in \chi_{AB} = (\chi_A, \chi_B)$ in the diatomic channel:

$$S_{AB}(\chi_{AB}|i) = -\sum_{j \in (A,B)} P(j|i) \log_2 P(j|i). \tag{228}$$

Finally, since in Eq. (227) the conditional entropy $S_{AB}(Y|\chi_X)$ quantifies (in bits) the X→Y delocalization per electron, so that the absolute IT-covalency in the diatomic fragment A—B reads:

$$\mathcal{S}_{AB} = N_{AB}\,[S_{AB}(B|\chi_A) + S_{AB}(A|\chi_B)]. \tag{229}$$

The bond-weighted contributions to the average *mutual*-information quantities (in bits) of the two bonded atoms are similarly defined in reference to the unbiased bond probabilities of AO [Eq. (226)]:

$$I_{AB}(\chi_A : B) = \sum_{i \in A} P_{AB}(i \wedge B)\ I(i : \chi_{AB}),$$
$$I_{AB}(A : \chi_B) = \sum_{i \in B} P_{AB}(i \wedge A)\ I(i : \chi_{AB}), \tag{230}$$

where:

$$I(\chi_{AB} : i) = \sum_{j \in (A,B)} P(j|i) \log_2 \left(\frac{P(j|i)}{p_b(j)} \right). \tag{231}$$

They generate the total IT-ionicity of all chemical bonds in the diatomic fragment AB [compare Eq. (229)]:

$$\mathcal{I}_{AB} = N_{AB}\,[I_{AB}(\chi_A : B) + I_{AB}(A : \chi_B)]. \tag{232}$$

Hence, the sum of the above total (diatomic) entropy-covalency and information-ionicity indices determines the overall information-theoretic bond multiplicity for the molecular fragment in question:

$$\mathcal{N}_{AB} = \mathcal{S}_{AB} + \mathcal{I}_{AB}. \tag{233}$$

Again, for diatomic molecules, for which $\chi_{AB} = \chi$ and the reference probabilities $\{p_b(k) = P(AB|k)/2 = ½\}$, the identity [see Eq. (215)]

$$\mathcal{N}_{AB} = \mathcal{S}_{AB} + \mathcal{I}_{AB} = \mathcal{M}(A, B) \tag{234}$$

can be readily demonstrated:

$$\begin{aligned}\mathcal{N}_{AB} &= \mathcal{S}_{AB} + \mathcal{I}_{AB}\\ &= N_{AB}\{\sum_{i\in A} P_{AB}(i\wedge B)\,[S_{AB}(\chi|i) + I(i:\chi)] + \sum_{i\in B} P_{AB}(i\wedge A)\,[S_{AB}(\chi|i) + I(i:\chi)]\}\\ &\equiv N_{AB}\{\sum_{i\in A} P_{AB}(i\wedge B) N(\chi;i) + \sum_{i\in B} P_{AB}(i\wedge A) N(\chi;i)\}\\ &= N_{AB}\{\sum_{i\in A} P_{AB}(i\wedge B) + \sum_{i\in B} P_{AB}(i\wedge A)\} = 2\, N_{AB}\, P_b(A\wedge B) = \mathcal{M}(A, B).\end{aligned} \tag{235}$$

Above, we have observed that the conditional IT bond multiplicity due to the input χ_k (per single electron)

$$\begin{aligned}N(\chi;k) &= \sum_{l\in\chi}\{-P(l|k)\log_2 P(l|k) + P(l|k)\log_2[P(l|k)/p_b(l)]\}\\ &= [\sum_{l\in\chi} P(l|k)]\,\log_2 2 = 1\,.\end{aligned} \tag{236}$$

In Table 4 we have compared the illustrative numerical RHF results [8b] of the IT bond multiplicities for the localized (diatomic) interactions in representative diatomic and polyatomic molecules for their equilibrium geometries. These predictions have been obtained using two choices of the Gaussian basis set: the minimum basis set (STO-3G), of the ground-state occupied Slater-type AO of the system constituent atoms, and the extended basis (6-31G*) including the Gaussian polarization functions.

Table 4. Comparison of the diatomic Wiberg index $\mathcal{W}$(A, B) and entropy/information bond-multiplicities $\mathcal{N}_{AB}$, S_{AB} and $\mathcal{I}_{AB}$ (in bits) from the bond-weighted AO communication channels in selected diatomic fragments A—B of representative molecules M: RHF results for equilibrium geometries in the minimum (STO-3G) and extended (6-31G*) basis sets (from [8b])

M	A–B	$\mathcal{W}$(A, B)		$\mathcal{N}_{AB}$		S_{AB}		$\mathcal{I}_{AB}$	
		Min.	Ext.	Min.	Ext.	Min.	Ext.	Min.	Ext.
F_2	F–F	1.000	1.228	1.000	1.228	0.947	1.014	0.053	0.273
HF	H–F	0.980	0.816	0.980	0.816	0.887	0.598	0.093	0.218
LiH	Li–H	1.000	1.005	1.000	1.005	0.997	1.002	0.003	0.004
LiF	Li–F	1.592	1.121	1.592	1.121	0.973	0.494	0.619	0.627
CO	C–O	2.605	2.904	2.605	2.904	2.094	2.371	0.511	0.533
H_2O	O–H	0.986	0.878	1.009	0.896	0.859	0.662	0.151	0.234
AlF_3	Al–F	1.071	1.147	1.093	1.154	0.781	0.748	0.311	0.406
CH_4	C–H	0.998	0.976	1.025	1.002	0.934	0.921	0.091	0.081
C_2H_6	C–C	1.023	1.129	1.069	1.184	0.998	1.078	0.071	0.106
	C–H	0.991	0.955	1.018	0.985	0.939	0.879	0.079	0.106
C_2H_4	C–C	2.028	2.162	2.086	2.226	1.999	2.118	0.087	0.108
	C–H	0.984	0.935	1.013	0.967	0.947	0.878	0.066	0.089
C_2H_2	C–C	3.003	3.128	3.063	3.192	2.980	3.095	0.062	0.097
	C–H	0.991	0.908	1.021	0.943	0.976	0.878	0.045	0.065
C_6H_6*	C_1–C_2	1.444	1.507	1.526	1.592	1.412	1.473	0.144	0.119
	C_1–C_3	0.00	0.061	0.000	0.059	0.000	0.035	0.000	0.024
	C_1–C_4	0.116	0.114	0.119	0.117	0.084	0.081	0.035	0.035

* For the sequential numbering of carbon atoms in the benzene ring.

In diatomic systems the trends exhibited by the entropic covalent and ionic components of the exactly conserved Wiberg overall bond order generally agree with intuitive chemical expectations. For example, in the minimum basis set description, the roughly "single" chemical bond in F_2, HF and LiH is seen to be almost purely covalent, although a more substantial IT-ionicity is diagnosed for the fluorine compounds in the extended basis set calculations. For the most ionic LiF, which exhibits in the minimum basis set roughly 3/2 bond, consisting of approximately 1 covalent and ½ ionic bond multiplicities, the extended basis set gives approximately a “single” bond-order estimate, with the information theory again predicting the ionic dominance over the covalent component of the resultant bond index. In CO, for which the extended basis set calculations have diagnosed approximately a “triple” bond, this chemical interaction is again seen to be predominantly covalent.

Table 5. The additional RHF predictions, obtained using the extended 6-31G* basis set (from [8b]) for illustrative diatomic and polyatomic molecules. The IT bond-orders are in bits.

Molecule	A-B	$\mathcal{M}$(A, B)	$\mathcal{N}_{AB}$	$\mathcal{S}_{AB}$	$\mathcal{I}_{AB}$
LiCl	Li-Cl	1.391	1.391	0.729	0.662
LiBr	Li-Br	1.394	1.394	0.732	0.662
NaF	Na-F	0.906	0.906	0.429	0.476
KF	K-F	0.834	0.834	0.371	0.463
SF_2	S-F	1.060	1.085	0.681	0.404
SF_4	S–F_a	1.055	1.064	0.670	0.394
	S-F_b	0.912	0.926	0.603	0.323
SF_6	S-F	0.978	0.979	0.726	0.254
B_2H_6†	B-B	0.823	0.851	0.787	0.063
	B-H_t	0.967	0.995	0.938	0.057
	B-H_b	0.476	0.490	0.462	0.028
Propellanes‡:					
[1.1.1]	C_b-C_b	0.797	0.829	0.757	0.072
[2.1.1]	C_b-C_b	0.827	0.860	0.794	0.066
[2.2.1]	C_b-C_b	0.946	0.986	0.874	0.112
[2.2.2]	C_b-C_b	1.009	1.049	0.986	0.063

‡ Central bonds between the *bridgehead* carbon atoms C_b.
† H_t and H_b denote the *terminal* and *bridge* hydrogen atoms, respectively.

The basis-set dependence of the predicted IT bond descriptors is seen to be relatively weak with the extended basis calculations often giving rise to predictions exhibiting slightly better agreement with intuitive chemical estimates. One also finds that in polyatomic systems the Wiberg bond-orders are very well reproduced by the overall IT descriptors. The carbon-carbon interactions in the benzene ring are seen to be properly differentiated and the intuitive multiplicities of the carbon-carbon chemical bonds in ethane, ethylene and acetylene are correctly accounted for.

The IT bond descriptors provide the covalent/ionic resolution of the Wiberg bond-order $\mathcal{M}$(A, B), which has been customarily regarded as being of purely "covalent" origin. However, the LCAO MO coefficients carry the information about both the electron-sharing (covalent) and electron-separation/transfer (ionic) phenomena in the chemical bond. Therefore, this overall index in fact combines the covalent and ionic contributions, which remain to be separated [46-52]. The present IT approach provides a novel resolution of this in fact *resultant* bond-order.

The significant information-ionicity contribution is also detected for all metal halides in the upper part of Table 5, where additional predictions from the extended basis set RHF calculations are reported. The subtle bond differentiation of the "equatorial" and "axial" S–F bonds in the irregular tetrahedron of SF_4 is correctly reproduced, and an increase in the strength of the central bond in propellanes with the increase in size of the bridges is correctly predicted.

Moreover, as intuitively expected, the C–H bonds are seen to slightly increase their information-ionicity, when the number of these terminal bonds increases in a series: acetylene, ethylene, ethane. In B_2H_6 the correct, around ½ bond-order of the bridging B–H bond is predicted and approximately single terminal bond multiplicity is detected. For the alkali metal fluorides the increase in the bond entropy covalency (decrease in information ionicity) with increasing size (softness) of the metal is also observed. For the fixed alkali metal in halides, e.g., in a series LiF, LiCl and LiBr, the overall bond order is increased for larger (softer) halogen atoms, mainly due to a higher entropy-covalency (delocalization, noise) component of the molecular communication channel in AO resolution.

6.6. Additive and Non-Additive Information Channels

Let us combine the molecular basis functions of typical LCAO MO calculations into the corresponding atomic subsets:

$$\chi = \{\chi_X\} = (\chi_A, \chi_B, \chi_C, \ldots) \equiv \chi^{AIM}. \qquad (237)$$

This arrangement determines the associated block structure of the AO conditional probability matrix:

$$\mathbf{P}(\chi^{\mathrm{AIM}}|\chi^{\mathrm{AIM}}) = \{\mathbf{P}(\chi_{\mathrm{X}}|\chi_{\mathrm{Y}})\}, \quad (\mathrm{X}, \mathrm{Y}) \in \mathrm{A}, \mathrm{B}, \mathrm{C}, \ldots \tag{238}$$

The diagonal block $\mathbf{P}(\chi_{\mathrm{X}}|\chi_{\mathrm{X}})$ determines the *internal* (*one*-center) communications $\mathrm{X}\rightarrow\mathrm{X}$ in atom X alone, which are responsible for the AIM *promotion* to its bonding (*valence*) state in the molecule. The *off*-diagonal blocks $\mathbf{P}(\chi_{\mathrm{X}}|\chi_{\mathrm{Y}})$ and $\mathbf{P}(\chi_{\mathrm{Y}}|\chi_{\mathrm{X}})$, $\mathrm{X} \neq \mathrm{Y}$, similarly generate the external (*two*-center) communications $\mathrm{Y}\rightarrow\mathrm{X}$ and $\mathrm{X}\rightarrow\mathrm{Y}$, respectively, between the contributed AO of both atoms, which are ultimately responsible for the truly *bonding* contributions to the overall IT multiplicities of the localized chemical bonds between the specified pairs of AIM. It should be emphasized, that the chemical values of diatomic bond multiplicities combine both the *one*- and *two*-center effects, of the *intra*-atom polarization (promotion) and *inter*-atomic delocalization and CT effects, respectively [7,8,46-52].

As we have already remarked in the preceding section, the *inter*-atomic communications in the molecular channel reflect the chemical interactions between atoms, so that the collection of *non*-bonded (separated) atoms of the promolecule exhibits only the *intra*-atomic probability propagations. The same principle can be used to naturally partition the molecular AO communication system of the AIM-arranged basis set χ^{AIM} into its *additive* and *non-additive sub*-channels [21]:

$$\mathbf{P}(\chi^{\mathrm{AIM}}|\chi^{\mathrm{AIM}}) \equiv \mathbf{P}^{total}(\chi^{\mathrm{AIM}}|\chi^{\mathrm{AIM}}) = \mathbf{P}^{add.}(\chi^{\mathrm{AIM}}|\chi^{\mathrm{AIM}}) + \mathbf{P}^{nadd.}(\chi^{\mathrm{AIM}}|\chi^{\mathrm{AIM}}). \tag{239}$$

As illustrated in Scheme 8, the former combines all *internal* (*intra*-atomic) communications within each (chemically decoupled) AIM, thus being solely determined by the diagonal, atomic blocks of the molecular conditional probabilities $\mathbf{P}(\chi^{\mathrm{AIM}}|\chi^{\mathrm{AIM}})$:

$$\mathbf{P}^{int.}(\chi^{\mathrm{AIM}}|\chi^{\mathrm{AIM}}) = \{\mathbf{P}(\chi_{\mathrm{X}}|\chi_{\mathrm{X}})\delta_{\mathrm{X,Y}}\} \equiv \mathbf{P}^{add.}(\chi^{\mathrm{AIM}}|\chi^{\mathrm{AIM}}). \tag{240}$$

The latter groups all complementary, *external* (*inter*-atomic) probability propagations between the (chemically coupled) pairs of bonded atoms in the molecular system under consideration:

$$\mathbf{P}^{ext.}(\chi^{AIM}|\chi^{AIM}) = \{\mathbf{P}(\chi_X|\chi_Y)(1 - \delta_{X,Y})\} \equiv \mathbf{P}^{nadd.}(\chi^{AIM}|\chi^{AIM}). \quad (241)$$

It should be stressed, however, that the *sub*-channel scattering probabilities originating from the given input do no longer sum up to 1, since this normalization condition applies only to the total list of outputs involving both the AIM diagonal and off-diagonal communications.

$$\begin{bmatrix} \mathbf{P}(\chi_A|\chi_A) & \mathbf{P}(\chi_B|\chi_A) & \mathbf{P}(\chi_C|\chi_A) & \cdots \\ \mathbf{P}(\chi_A|\chi_B) & \mathbf{P}(\chi_B|\chi_B) & \mathbf{P}(\chi_C|\chi_B) & \cdots \\ \mathbf{P}(\chi_A|\chi_C) & \mathbf{P}(\chi_B|\chi_C) & \mathbf{P}(\chi_C|\chi_C) & \cdots \\ \cdots & \cdots & \cdots & \cdots \end{bmatrix}$$

$$= \begin{bmatrix} \mathbf{P}(\chi_A|\chi_A) & \mathbf{0} & \mathbf{0} & \cdots \\ \mathbf{0} & \mathbf{P}(\chi_B|\chi_B) & \mathbf{0} & \cdots \\ \mathbf{0} & \mathbf{0} & \mathbf{P}(\chi_C|\chi_C) & \cdots \\ \cdots & \cdots & \cdots & \cdots \end{bmatrix} + \begin{bmatrix} \mathbf{0} & \mathbf{P}(\chi_B|\chi_A) & \mathbf{P}(\chi_C|\chi_A) & \cdots \\ \mathbf{P}(\chi_A|\chi_B) & \mathbf{0} & \mathbf{P}(\chi_C|\chi_B) & \cdots \\ \mathbf{P}(\chi_A|\chi_C) & \mathbf{P}(\chi_B|\chi_C) & \mathbf{0} & \cdots \\ \cdots & \cdots & \cdots & \cdots \end{bmatrix}$$

$$= \begin{bmatrix} A \to A & \mathbf{0} & \mathbf{0} & \cdots \\ \mathbf{0} & B \to B & \mathbf{0} & \cdots \\ \mathbf{0} & \mathbf{0} & C \to C & \cdots \\ \cdots & \cdots & \cdots & \cdots \end{bmatrix} + \begin{bmatrix} \mathbf{0} & A \to B & A \to C & \cdots \\ B \to A & \mathbf{0} & B \to B & \cdots \\ C \to A & C \to B & \mathbf{0} & \cdots \\ \cdots & \cdots & \cdots & \cdots \end{bmatrix}$$

total = internal (additive) + external (*non*-additive)

Scheme 8. Partitioning of the conditional AO probabilities defining the molecular information system in this resolution into the *one*-center (AIM-internal, additive) and *two*-center (AIM-external, *non*-additive) *sub*-channels, and the underlying communications between the constituent bonded atoms

In CTCB/OCT approaches the *full* list of the AIM/AO inputs determines the complete *origins* (sources) of all chemical bonds in the molecule, while the full list of such outputs signifies that all chemical bonds are being counted in the resulting bond-multiplicity descriptors. Thus, should one focus on the effective chemical bonds between the specified pair of atoms A and B, only the χ_{AB} outputs should be included in the relevant communication network. Again, the full list χ

of inputs then generates the resultant chemical connectivity between the two bonded atoms in the molecule under consideration, while limiting this list to χ_{AB} generates corresponding entropy/information measures due to internal communications in this diatomic fragment. Therefore, both the *intra*-atom and *inter*-atom communications ultimately contribute to the overall IT bond index in the molecular system in question. Indeed, the chemical bond concept combines both the atom promotion (polarization) and the inter-atomic delocalization/CT phenomenona.As we have already argued in the preceding section, the promoted (valence) state of each AIM is determined mainly by the associated atomic (diagonal) block of molecular conditional probabilities. Important though it is for the full characterization of the AIM valence preparation in the molecule and the resultant, chemical bond multiplicities, which are reflected by the resultant IT bond-multiplicities and their covalent/ionic components, it has no direct relevance for the pattern of diatomic interactions between bonded atoms. Therefore, the partition of Eqs (239) - (241) again emphasizes the importance of separating the additive and *non*-additive *sub*-channels, for distinguishing the chemical *one*-center promotion of AIM from the *two*-center interaction phenomena in OCT of the chemical bond.

Clearly, in the canonical AO representation there are always some *internal* covalency ("noise") and ionicity (information flow) contributions involved in the atomic promotion processes. One also observes that in the *Natural Hybrid Orbital* (NHO) freamework, in which the atomic (diagonal) blocks of the *first*-order density matrix become diagonal themselves, the *intra*-atomic communications become deterministic in character, so that the *one*-center (additive) IT-covalency identically vanishes. As in the CG probe of the chemical bond localization, the entropic bond descriptors of diatomic interactions are then seen to be solely determined by the *non*-additive *sub*-channel, which combines the communications between AO originating from *different* atoms. When supplemented by the atom-promotion communications and the ensemble bond-weighting in the channel input, this separation exactly reproduces the Wiberg bond-orders in diatomic molecules (See Section 6.5) and fully accounts for the chemical bond-differentiation pattern in diatomic fragments of typical molecules.

The partitioning of the conditional probabilities in Eqs. (239)-(241) is in the spirit of a related division of the AO-representations of quantum-mechanical operators in a similar context of their *internal* and *external* eigenvalue problems [71,72,76]. The latter approach has been successfully applied in identifying the partially-decoupled channels of the collective electron displacements in reactants [71,72], and it has been used to determine the *inter*-atomic flows of electrons in molecules [77-79].

6.7. *Many*-Orbital Phenomena in OCT

The entropy/information quantities for several (dependent) probability schemes (see Section 1.5), based upon the coupling measures of the joint probabilities of *many*-orbital events in the molecular bond system (Appendix D), have recently been used to describe the inter-fragment IT couplings between internal orbital communications in molecular subsystems [22]. In the remaining part of this section we shall briefly examine potential applications of these generalized bond descriptors in probing subtle bonding effects in molecular and/or reactive systems and their fragments. For simplicity, we shall limit out discussion to IT quantities involving three or four probability schemes in the AO resolution (see Schemes 4 and 5). The molecular scenarios invoking IT quantities of three probability distributions may involve three separate species A, B and C, e.g., two reactants A and B and the catalyst/surface C, with the corresponding sets of the AO-events ($\boldsymbol{a}$, $\boldsymbol{b}$, $\boldsymbol{c}$) of the associated probability schemes (**A**, **B**, **C**) then referring to the basis functions provided by the constituent atoms of each subsystem. Alternatively, three molecular fragments can be involved. The *three*-orbital development then enables one to discuss the influence of one reactant/fragment, say C, on the bond structure (or reactivity) of two remaining fragments A and B. For example, one could address in such IT framework a natural question about the influence of the catalyst on the structure/reactivity of two adsorbed species, and ultimately assess the cooperation effects between the catalyst-adsorbate bonds (A—C, B—C) and the A—B bond linking the two adsorbates. For this catalytic scenario the conditional *three*-scheme entropy

$$S(\boldsymbol{P}(\boldsymbol{b})|\mathbf{P}(\boldsymbol{a}\wedge\boldsymbol{c})) = H(\mathbf{B}|\mathbf{AC}) = H(\mathbf{B}|\mathbf{A}) - I(\mathbf{B}{:}\mathbf{C}|\mathbf{A}), \tag{242}$$

which reflects the molecular indeterminacy of the AO-output events $\boldsymbol{b}$ in B with respect to the input product-events $\boldsymbol{a}\wedge\boldsymbol{c}$ in the combined subsystem (A¦C). As seen in Scheme 4, this IT-covalency accounts for only a *part* of the overall noise component in the $\boldsymbol{a}\to\boldsymbol{b}$ communications measured by the indeterminacy of $\boldsymbol{b}$ with respect to $\boldsymbol{a}$ alone, $H(\mathbf{B}|\mathbf{A}) = S(\boldsymbol{P}(\boldsymbol{b})|\boldsymbol{P}(\boldsymbol{a}))$, which reflects the *overall* information loss in the $\boldsymbol{a}\to\boldsymbol{b}$ probability scattering in between A and B. The remaining part, the mutual-information $I(\mathbf{B}{:}\mathbf{C}|\mathbf{A})$, by which $H(\mathbf{B}|\mathbf{AC})$ is decreased relative to $H(\mathbf{B}|\mathbf{A})$ and by which $I(\mathbf{B}{:}\mathbf{AC}) = I(\mathbf{A}{:}\mathbf{B}) + I(\mathbf{B}{:}\mathbf{C}|\mathbf{A})$ is increased relative to $I(\mathbf{A}{:}\mathbf{B})$, should now be attributed to the influence of the catalyst C in interactions between B and (A¦C). Similarly, $H(\mathbf{A}|\mathbf{BC}) = H(\mathbf{A}|\mathbf{B}) - I(\mathbf{A}{:}\mathbf{C}|\mathbf{B})$ and $I(\mathbf{A}{:}\mathbf{BC}) = I(\mathbf{A}{:}\mathbf{B}) + I(\mathbf{A}{:}\mathbf{C}|\mathbf{B})$, so that again the presence of the catalyst C is felt by an effective increase in IT-ionic character of interaction of A with (B¦C) compared to the separate interaction of A with B alone. Thus, the mutual information in **A** and **C**, given **B**, $I(\mathbf{A}:\mathbf{C}|\mathbf{B})$, and in **B** and **C**, given **A**, $I(\mathbf{B}:\mathbf{C}|\mathbf{A})$, together account for the effect of an increased ionic (deterministic) character of the chemical interactions between the chemisorbed reactants A and B, compared to the AO communications involving separate species:

$$I(\mathbf{A}:\mathbf{BC}) + I(\mathbf{B}:\mathbf{AC}) - 2I(\mathbf{A}:\mathbf{B}) = I(\mathbf{A}:\mathbf{C}|\mathbf{B}) + I(\mathbf{B}:\mathbf{C}|\mathbf{A}). \quad (243)$$

This increase in A—B IT-ionicity (decrease in A—B IT covalency) is due to the fact that due to the electron delocalization Y→C the chemisorbed reactant Y which now represents the modified interaction partner of X in the combined subsystem (Y¦C), effectively increases its entropy circle by an extra area (uncertainty) provided by the catalyst in the (Y,C) envelope. The above IT-ionic "activation" of adsorbates [20], as a result of forming the partial A—C and B—C bonds on active sites of the catalytic surface, also manifests the *competition effect* between these surface bonds and the *inter*-adsorbate bond A—B in the catalytic system: the more heavily are the valence electrons of A and B involved in chemical bonds with C, the less noisy (more deterministic) are their mutual communications, thus giving rise to less IT-covalent (more IT-ionic) interactions between the *chemisorbed* species. The *physical* adsorption of these reactants should be marked by a relatively small value of $I(\mathbf{B}:\mathbf{C}|\mathbf{A})$, since then the dependencies ("overlaps") between entropy circles of schemes (**B**, **A**) and **C**, should be relatively small, thus grossly diminishing the above IT-ionic activation effect generated by the presence of the catalyst. The surface *chemical* bond between a given adsorbate and the catalyst should be strongly felt at the position of the other adsorbate, and hence the information "coupling" between probability distributions of the chemisorbed species and that of the catalyst should be relatively strong.

Another molecular scenario, in which the *three*-scheme entropy/information descriptors are expected to be useful, is the influence of one reactive site in a molecule upon another, e.g., in the contexts of subtle reactivity preferences in the *donor-acceptor* (DA) reactive systems implied by the *Hard and Soft Acids and Bases* (HSAB) rule [66-72], and particularly - its regional formulation for predicting the regioselectivity trends in cyclization reactions [73], the *Maximum Complementarity* principle [74], and the *Bond-Length Variation Rules* of Gutmann [75]. Such cooperative interaction between different sites in a molecule is also responsible for the directing *trans/cis influence* of ligands in transition metal complexes and the familiar *substituent effect* in aromatic systems. The adequate IT description of the AIM cooperation in *many*-centre bonds, e.g., in boron hydrides or propellanes, may also require the entropy/information indices involving several probability schemes.

The four probability schemes generate a diversity of the conditional-entropy and mutual-information descriptors introduced in Section 1.5. They have been delineated in the qualitative entropy diagrams of Schemes 5, 9 and 10. Scheme 9 and 10 correspond to weakly and strongly (chemically) interacting subsystems M_1 and M_2 in $M = (M_1 \vert M_2)$, which involve smaller molecular fragments of interest, M_1 = A—B and M_2 = C—D, to which the probability schemes **A**, **B**, **C**, **D** are ascribed. Alternatively, the pairs of the input and output distributions in each sysbsystem may constitute the four schemes under consideration. Let us examine the IT-covalent and IT-ionic couplings between the orbital information systems of these complementary subsystems in M (see Scheme 11) [20,22]. Let schemes **A** and **B** respectively denote the AO input and outputs in M_1 with the remaining schemes **C** and **D** having a similar meaning for M_2. It should be stressed that each scheme in M_1 now combines the AO contributed by its both constituent fragments A and B, while each scheme in M_2 combines the AO of its fragments C and D, with the fragment AO events extending over all basis functions contributed by its constituent atoms. Our aim now is to identify prospective candidates for the entropy-information descriptors of the mutual influence of chemical bond(s) in M_1, originating from communications $\boldsymbol{a \to b}$, on bonds in M_2, generated by communications $\boldsymbol{c \to d}$. These *intra*-fragment communications are characterized by the conditional probabilities $\mathbf{P(B|A)} = \{P(b|a)\}$ and $\mathbf{P(D|C)} = \{P(d|c)\}$, respectively, which define the associated *singly*-conditional probability schemes **(B|A)** and **(D|C)**. The mutual dependencies of the internal communications in the complementary subsystems of M are then described by the *triply*-conditional probabilities of the probability propagations $\boldsymbol{a \to b}$ in M_1 conditional on the information scattering $\boldsymbol{c \to d}$ in M_2:

$$\mathbf{P(B|A \| D|C)} = \{P(b|a \| d|c) = P[(b|a) \wedge (d|c)]/P(d|c)\}. \tag{244}$$

Here the joint probabilities of two conditional AO events, $P[(b|a) \wedge (d|c)]$ satisfy the usual normalization conditions,

$$\sum_{(b|a)}^{M_1} \sum_{(d|c)}^{M_2} P[(b|a) \wedge (d|c)] = \sum_{(d|c)}^{M_2} P(d|c) = 1 \cdot \tag{245}$$

with summations ranging over all internal communications in each subsystem.

The average conditional-entropy indices of this *inter*-fragment covalent coupling read:

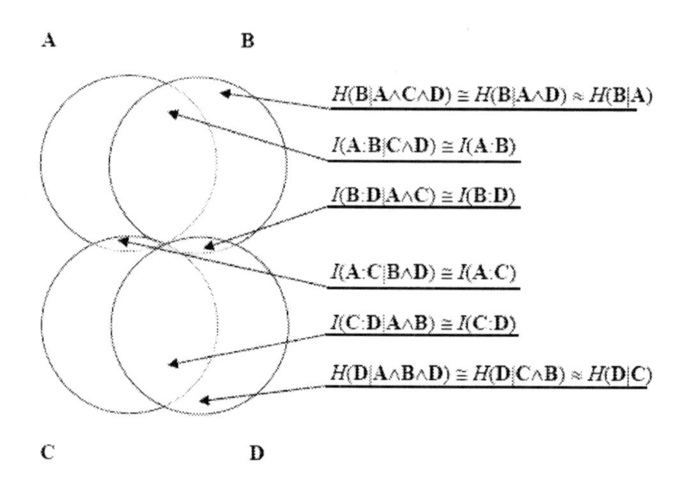

Scheme 9. General entropy/information diagrams of four probability schemes (**A, B, C, D**) corresponding to the weakly-interacting, internally-bonded subsystems A—B and C—D in the reaction complex $\begin{bmatrix} \mathrm{A—B} \\ \vdots \quad \vdots \\ \mathrm{C—D} \end{bmatrix}$ (compare Scheme 5).

$$H(\mathbf{B}|\mathbf{A}\,\|\,\mathbf{D}|\mathbf{C}) = -\sum_{(b|a)}^{\mathrm{M}_1}\sum_{(d|c)}^{\mathrm{M}_2} P[(b|a)\wedge(d|c)]\log P(b|a\|d|c),$$

$$H(\mathbf{D}|\mathbf{C}\,\|\,\mathbf{B}|\mathbf{A}) = -\sum_{(b|a)}^{\mathrm{M}_1}\sum_{(d|c)}^{\mathrm{M}_2} P[(b|a)\wedge(d|c)]\log P(d|c\|b|a), \tag{246}$$

while the IT-ionic coupling between the two fragments is embodied in the associated average mutual information quantity:

$$I[(\mathbf{B}|\mathbf{A}):(\mathbf{D}|\mathbf{C})] = \sum_{(b|a)}^{\mathrm{M}_1}\sum_{(d|c)}^{\mathrm{M}_2} P[(b|a)\wedge(d|c)]\log\frac{P[(b|a)\wedge(d|c)]}{P(b|a)\,P(d|c)}$$

$$= H(\mathbf{B}|\mathbf{A}) - H(\mathbf{B}|\mathbf{A}\,\|\,\mathbf{D}|\mathbf{C}) = H(\mathbf{D}|\mathbf{C}) - H(\mathbf{D}|\mathbf{C}\,\|\,\mathbf{B}|\mathbf{A}). \tag{247}$$

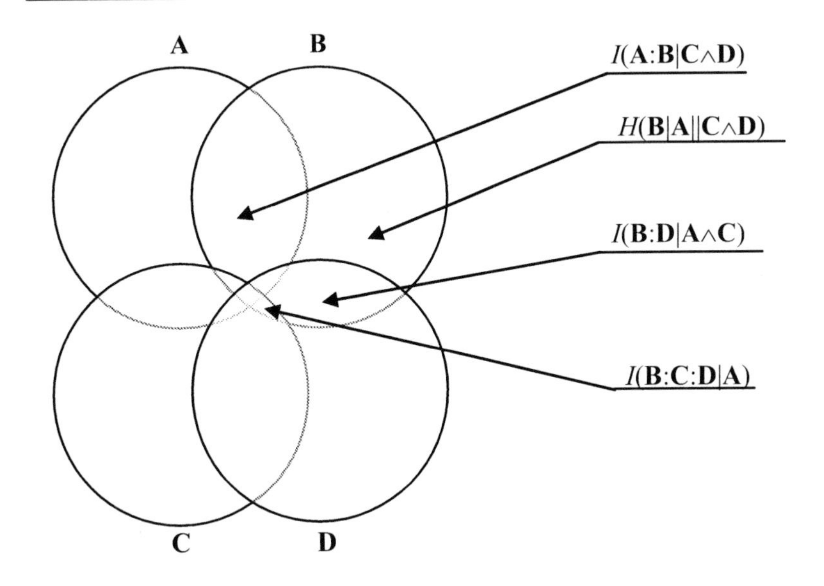

Scheme 10. Entropy/information diagram of the interaction between the internal communications ***a*→*b*** and ***c*→*d*** in the strongly interacting complementary fragments $M_1 = (A{\mid}B)$ and $M_2 = (C{\mid}D)$ of $M = (M_1{\mid}M_2)$.

These IT covalency and ionicity descriptors of the communication interaction between the two subsystems thus conserve the internal uncertainties in each molecular fragment (see Scheme 11) measured by the corresponding Shannon entropies of the internal communications in each subsystem:

$$N[(\mathbf{B}|\mathbf{A});(\mathbf{D}|\mathbf{C})] = H(\mathbf{B}|\mathbf{A}\,\|\,\mathbf{D}|\mathbf{C}) + I[(\mathbf{B}|\mathbf{A}):(\mathbf{D}|\mathbf{C})] = H(\mathbf{B}|\mathbf{A}),$$

$$N[(\mathbf{D}|\mathbf{C});(\mathbf{B}|\mathbf{A})] = H(\mathbf{D}|\mathbf{C}\,\|\,\mathbf{B}|\mathbf{A}) + I[(\mathbf{B}|\mathbf{A}):(\mathbf{D}|\mathbf{C})] = H(\mathbf{D}|\mathbf{C}). \tag{248}$$

Let us now ascribe the four probability schemes (**A, B, C, D**) to AO-events in fragments A, B (of M_1) and C, D (of M_2), respectively (see Scheme 10). Other entropy/information descriptors discussed in Section 1.5 should also reflect specific chemical influences between chemically interacting subsystems in $M = (M_1{\mid}M_2)$. For example, the mutual-information $I(\mathbf{A}:\mathbf{B}\,|\,\mathbf{C}\wedge\mathbf{D})$ [Eq. (48)] accounts for the influence of M_2 on the IT-ionicity in M_1. The associated conditional-entropy,

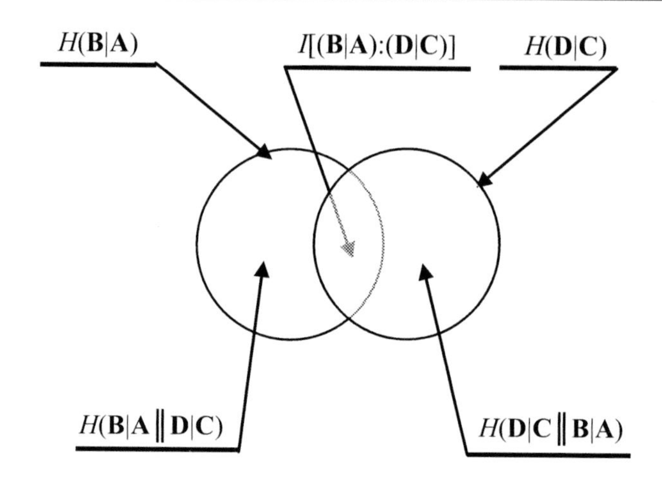

Scheme 11. Entropy/information diagram of the interaction between the internal communications $\boldsymbol{a}\rightarrow\boldsymbol{b}$ and $\boldsymbol{c}\rightarrow\boldsymbol{d}$ in the complementary fragments M_1 and M_2 of $M = (M_1 \vert M_2)$.

$$H(\mathbf{B}|\mathbf{A} \,\|\, \mathbf{C}\wedge\mathbf{D}) = H(\mathbf{B}|\mathbf{A}) - I(\mathbf{B}{:}\mathbf{D}|\mathbf{A}\wedge\mathbf{C}) - I(\mathbf{B}{:}\mathbf{C}{:}\mathbf{D}|\mathbf{A}), \tag{249}$$

similarly reflects the effect of M_2 on the IT-covalency in M_1.

Finally, by attributing the four schemes **A**, **B**, **C** and **D** to the corresponding fragments in the weakly interacting bimolecular reactive system (see Scheme 9),

$$M_1\text{----}M_2 = \begin{bmatrix} A\text{—}B \\ \vdots \quad \vdots \\ C\text{—}D \end{bmatrix}, \tag{250}$$

one predicts $I(\mathbf{A}{:}\mathbf{B} \,|\, \mathbf{C}\wedge\mathbf{D}) \cong I(\mathbf{A}{:}\mathbf{B})$, $I(\mathbf{B}{:}\mathbf{C}{:}\mathbf{D} \,|\, \mathbf{A}) \cong 0$, and hence $H(\mathbf{B}|\mathbf{A} \,\|\, \mathbf{C}\wedge\mathbf{D}) \cong H(\mathbf{B}|\mathbf{A}) - I(\mathbf{B}{:}\mathbf{D}|\mathbf{A}\wedge\mathbf{C}) \approx H(\mathbf{B}|\mathbf{A})$. Therefore, the weak M_1----M_2 interactions have practically vanishing effect on the internal ionicities of reactants, with only their internal covalencies being slightly affected.

CONCLUSION

In this monograph we have introduced the key concepts and techniques of IT, which have subsequently been used to explore the electronic structure of molecules. The recent progress and representative applications of such IT of molecular systems has been surveyed. Among other developments the use of information densities as local probes of the equilibrium electronic distributions and of the structure of bonded atoms has been emphasized. Information theory also provides a novel perspective on both the location and the entropic origins of the chemical bonds. The Fisher information principle has been used to generate the Schödinger equations of quantum mechanics and the variational principle of the entropy-deficiency, relative to the atomic promolecule, has been shown to give rise to the IT-unbiased, Hirshfeld scheme of partitioning the molecular electron density into the infinitely-extending and the subsystem v-representable "stockholder" atoms. The importance of the *non*-additive entropy/information measures in extracting subtle changes due to the chemical bond formation has been stressed. In particular the densities of the *non*-additive Fisher information in the MO and AO resolutions, which emphasize the importance of the electronic kinetic energy in the bond-formation phenomena, have been advocated as useful electron localization (ELF) and bond detection (CG) probes, respectively. Illustrative numerical results have been presented to validate this claim.

It should be emphasized, however, that the "*vertical*", *sub*-molecular reality of the subsystem resolution [7], which is so important for the "language" and understanding in chemistry, cannot be established by a direct experiment, since AIM cannot be formulated as the unique quantum-mechanical "observables". Therefore, the bonded atoms, functional groups and chemical bonds have to be ultimately classified as *Kantian noumenons* [15]. Nonetheless, the partial understanding and indirect probes of these important chemical concepts are available from several different perspectives. The close analogy between the phenomenological description of molecules and their fragments in IT and the ordinary thermodynamics further validates these chemical constructs, since it introduces the thermodynamic-like causality into relations between perturbations and responses of molecular subsystems and hence brings more consistency into the "theory" of chemistry [7].

Of similar character are recent developments in CTCB, and particularly in OCT, which introduce a novel communication (entropy/information) perspective on several classical issues in the electronic structure theory. Until recently, a wider use of CTCB [7] in probing the molecular electronic structure has been hindered by the originally *two*-electron conditional probabilities, which blur a diversity of chemical bonds. The AO-resolved OCT using the flexible-input probabilities and recognizing the bonding/antibonding character of the orbital interactions in a molecule, reflected by the signs of the underlying CBO matrix elements, to a large extent remedies this problem [8]. The off-diagonal conditional probabilities it generates are proportional to the quadratic bond indices of the MO theory, and hence the strong inter-orbital communications correspond to strong Wiberg bond-orders. It also properly accounts for the increasing populational decoupling of AO, when the anti-bonding MO become occupied [8b]. The diatomic development in OCT extends our understanding of the chemical bond from the complementary viewpoint of the Communication Theory. The bond-probability weighting of contributions due to separate AO inputs gives excellent results, which fully reproduce the bond differentiation in diatomic fragments of the molecule, as manifested by the quadratic bond index of Wiberg, and gives rise to its resolution into the associated IT-covalent and IT-ionic components. We have also briefly outlined the many-orbital generalization of OCT, which allows one to describe the inter-bond coupling phenomena and opens a novel, IT-perspective on the origins of catalytical activity and multi-bond reactivity.

It should be also emphasized, that the extra computation effort of the complementary IT analysis of the molecular bonding patterns is negligible, compared to the cost of standard SCF LCAO MO calculations of the molecular electronic structure, since practically all quantum-mechanical computations in the orbital approximation already determine both the kinetic-energy and the CBO data required, e.g., by CG analysis and OCT.

The IT approach is very much in spirit of the Eugene Wigner's observation, often quoted by Walter Kohn, that the understanding in science requires insights from several different points of view. The kinetic-energy probe of the bonding regions in molecules and the information perspective on the genesis of the chemical bond provide such an alternative. Only together these complementary tools constitute what one would call a more "complete" theory of the complex bond phenomenon, which – to paraphrase yet another famous citation from Samuel Beckett – is one of old, good problems that never die out.

Appendix A

INFORMATION CONTINUITY REVISITED

A.1. PROBABILITY, CURRENT AND INFORMATION DENSITIES

In the stationary quantum state of N electrons,

$$\Psi(N) = \psi(\{\boldsymbol{r}_k\}) \exp(-iEt/\hbar) \equiv \psi(\{\boldsymbol{r}_k\}) \exp(-i\omega t), \tag{A.1}$$

which conserves the system energy E, the probability-current density,

$$\boldsymbol{j}(\boldsymbol{r}) = \langle \Psi(N) | \hat{\mathbf{j}}(\boldsymbol{r}) | \Psi(N) \rangle, \tag{A.2}$$

where

$$\hat{\mathbf{j}}(\boldsymbol{r}) = \frac{\hbar}{2m_e i} \sum_{k=1}^{N} [\nabla_k \delta(\boldsymbol{r}_k - \boldsymbol{r}) + \delta(\boldsymbol{r}_k - \boldsymbol{r}) \nabla_k], \tag{A.3}$$

with m_e standing for the electronic mass, and the associated current contribution to the Fisher-information both identically vanish for the real spatial (amplitude) function $\psi(\{\boldsymbol{r}_k\})$. Indeed, in such states the wave function exhibits the (energy dependent) time phase-factor, which does not depend on particle positions. For the complex amplitude factor in the stationary wave function, however, e.g., in degenerate electronic states, and in a general case of the non-stationary state the finite probability-current and the associated information-current play a vital role in the time evolution of the probability/information distributions in molecular systems.

Let us again explicitly consider the simplest case of a single (spin-less) particle of mass $\mu = m_e$ in the system characterized by the (Hermitian) Hamiltonian operator

$$\hat{\mathrm{H}}(\boldsymbol{r}) = -(\hbar^2/2m_e)\nabla^2 + v(\boldsymbol{r}) = \hat{\mathrm{T}}_e(\boldsymbol{r}) + \hat{\mathrm{v}}(\boldsymbol{r}), \tag{A.4}$$

with the multiplicative operator $\hat{\mathrm{v}}(\boldsymbol{r}) = v(\boldsymbol{r})$ corresponding to the external potential due to the system nuclei in their fixed positions, in accordance with the familiar Born-Oppenheimer (adiabatic) approach to molecular electronic structure. The particle quantum-mechanical state, described by the (complex) wave function

$$\psi(\boldsymbol{r}, t) = R(\boldsymbol{r}, t)\exp[i\Theta(\boldsymbol{r}, t)], \tag{A.5}$$

where the real functions $R(\boldsymbol{r}, t)$ and $\Theta(\boldsymbol{r}, t)$ describe the probability-amplitude and the phase of ψ, respectively, gives rise to the probability density,

$$p(\boldsymbol{r}, t) = |\psi(\boldsymbol{r}, t)|^2 = R(\boldsymbol{r}, t)^2, \qquad \int p(\boldsymbol{r}, t)\, d\boldsymbol{r} = 1, \tag{A.6}$$

and the probability-current density:

$$\boldsymbol{j}(\boldsymbol{r}, t) = \frac{\hbar}{2m_e i}(\psi^*\nabla\psi - \psi\nabla\psi^*) = \frac{\hbar}{m_e}\mathrm{Im}(\psi^*\nabla\psi)$$

$$= \frac{\hbar}{m_e} p\nabla\Theta = p\nabla\left[\frac{\hbar\Theta}{m_e}\right] \equiv p(\boldsymbol{r}, t)\boldsymbol{V}(\boldsymbol{r}, t). \tag{A.7}$$

The preceding equation also expresses the local velocity of the probability "fluid", $\boldsymbol{V}(\boldsymbol{r}, t) = \boldsymbol{j}(\boldsymbol{r}, t)/p(\boldsymbol{r}, t)$, in terms of the phase-part of the wave function.

For simplicity, in what follows the factor 4 in the amplitude definition of the Fisher information [Eq. (6)] is removed by explicitly considering the renormalized gradient measure,

$$\bar{I}[\psi] = I[\psi]/4 \equiv \int \bar{f}(\boldsymbol{r})\, d\boldsymbol{r}, \tag{A.8}$$

determined by its density

$$\bar{f}(\boldsymbol{r}) = \nabla\psi^*(\boldsymbol{r})\cdot\nabla\psi(\boldsymbol{r}). \qquad \text{(A.9)}$$

In terms of the amplitude and phase parts of the complex quantum state of Eq. (A5) this Fisher information reads:

$$\bar{I}[\psi] = \int[(\nabla R)^2 + p(\nabla\Theta)^2]\,d\boldsymbol{r} = \frac{1}{4}\int\frac{(\nabla p)^2}{p}d\boldsymbol{r} + \frac{m_e^2}{\hbar^2}\int\frac{\boldsymbol{j}^2}{p}d\boldsymbol{r}$$

$$\equiv \int\bar{f}_p(\boldsymbol{r})d\boldsymbol{r} + \int\bar{f}_j(\boldsymbol{r})d\boldsymbol{r} \equiv \bar{I}[p,\boldsymbol{j}]. \qquad \text{(A.10)}$$

Therefore, both the probability distribution p and the paramagnetic current density $\boldsymbol{j}$ contribute to the resultant integral information content $\bar{I}[\psi]$ of the system wave function and to its density $\bar{f}[\psi;\boldsymbol{r}] = \bar{f}_p(\boldsymbol{r}) + \bar{f}_j(\boldsymbol{r})$. In other words, the probability distribution and the current density determine the two independent degrees of freedom of the system information content in molecular systems. The associated expression for the linear displacement in the overall information content reads:

$$\delta\bar{I}[p,\boldsymbol{j}] = \int\left[\frac{1}{4}\left(\frac{\nabla p(\boldsymbol{r})}{p(\boldsymbol{r})}\right)^2 - \frac{1}{2}\frac{\nabla^2 p(\boldsymbol{r})}{p(\boldsymbol{r})}\right]\delta p(\boldsymbol{r})d\boldsymbol{r}$$

$$+\left(\frac{m_e}{\hbar p}\right)^2\int[2\boldsymbol{j}(\boldsymbol{r})\cdot\delta\boldsymbol{j}(\boldsymbol{r}) - \boldsymbol{j}^2(\boldsymbol{r})\delta p(\boldsymbol{r})]d\boldsymbol{r}. \qquad \text{(A.11)}$$

A.2. TIME-DEPENDENT PHEMOMENA AND STATIONARY ACTION PRINCIPLE

Time evolution of the above single-particle wave function is determined by the *non*-stationary Schrödinger equation:

$$i\hbar\frac{\partial\psi}{\partial t} = \hat{\mathrm{H}}\psi \qquad \text{or} \qquad \left(i\hbar\frac{\partial}{\partial t} - \hat{\mathrm{H}}\right)\psi \equiv \hat{\mathrm{A}}\psi = 0, \qquad \text{(A.12)}$$

where $\hat{\mathrm{A}}(\boldsymbol{r},t)$ denotes the operator determining the quantum mechanical action integral:

$$\mathcal{A} = \int_{t_0}^{t_1} dt \langle \Psi(t) | \hat{\mathrm{A}}(t) | \Psi(t) \rangle \cdot \tag{A.13}$$

Equation (A.12) is seen to result from the *stationary-action principle* $\delta\mathcal{A} = 0$ or

$$\frac{\delta\mathcal{A}}{\delta\psi^*(\boldsymbol{r},t)} = 0 , \tag{A.14}$$

which replaces in the time-dependent problem the energy variational principle valid for the stationary case, when the system energy is conserved. It implies the conservation in time of the wave-function (probability) normalization:

$$\frac{d}{dt}\int \psi^*(\boldsymbol{r},t)\,\psi(\boldsymbol{r},t)\,d\boldsymbol{r} = \frac{d}{dt}\int p(\boldsymbol{r},t)\,d\boldsymbol{r} = 0 \,. \tag{A.15}$$

In the *Time-Dependent Density Functional Theory* (TDDFT) [80,81], for the given initial state $\psi(\boldsymbol{r}, t_0)$ the time-dependent density determines the external potential uniquely up to within an additive, purely time-dependent function. The latter in turn determines the time dependent wave-function up to within a purely time-dependent phase: $\Psi(t) = \exp[-i\alpha(t)]\,\tilde{\Psi}[\rho;t]$. This phase ambiguity cancels out in the expectation values of physical quantities, which for the given initial state can be thus regarded as the unique functionals of the system time-dependent probability distribution $p(\boldsymbol{r}, t)$, or of the associated electron density $\rho(\boldsymbol{r}, t) = Np(\boldsymbol{r}, t)$, e.g.,

$$j(\boldsymbol{r}, t) = \langle \Psi(t) | \hat{\mathrm{j}}(\boldsymbol{r}) | \Psi(t) \rangle = j[\rho; \boldsymbol{r}, t]. \tag{A.16}$$

From the familiar expression for the time-dependence of the quantum-mechanical expectation values,

$$\frac{\partial}{\partial t}\langle\Psi(t)|\hat{Q}(t)|\Psi(t)\rangle = \langle\Psi(t)|\frac{\partial\hat{Q}(t)}{\partial t} - \frac{i}{\hbar}[\hat{Q}(t),\hat{H}(t)]|\Psi(t)\rangle, \qquad (A.17)$$

one then obtains for $\hat{Q}(\boldsymbol{r},t) = \hat{\mathbf{j}}(\boldsymbol{r})$:

$$\partial \boldsymbol{j}(\boldsymbol{r},t)/\partial t = \partial\langle\Psi(t)|\hat{\mathbf{j}}(\boldsymbol{r})|\Psi(t)\rangle/\partial t$$

$$= \frac{i}{\hbar}\langle\Psi(t)|[\hat{H}(t),\hat{\mathbf{j}}(\boldsymbol{r})]|\Psi(t)\rangle \equiv \boldsymbol{G}[\rho;\boldsymbol{r},t]. \qquad (A.18)$$

It should be finally observed that the information density of Eq. (A.9) represents the expectation value

$$\bar{f}(\boldsymbol{r}) = \nabla\psi^*(\boldsymbol{r})\cdot\nabla\psi(\boldsymbol{r}) = \langle\psi(t)|\hat{\bar{\mathrm{f}}}(\boldsymbol{r})|\psi(t)\rangle, \qquad (A.19)$$

of the associated information-density operator:

$$\hat{\bar{\mathrm{f}}}(\boldsymbol{r}) = \frac{\hat{\mathbf{p}}^2(\boldsymbol{r})}{\hbar^2} \equiv \hat{\mathbf{k}}^2(\boldsymbol{r}) = -\Delta_{r'}\delta(\boldsymbol{r}'-\boldsymbol{r}), \qquad (A.20)$$

where the particle momentum operator $\hat{\mathbf{p}}(\boldsymbol{r}) \equiv \hbar\hat{\mathbf{k}}(\boldsymbol{r}) = -i\hbar\nabla_r$. Indeed, the expectation value of Eq. (A.19) then recovers the known amplitude form of the quantum-mechanical information density:

$$\bar{f}(\boldsymbol{r}) = -\int\psi^*(\boldsymbol{r}',t)\Delta_{r'}\psi(\boldsymbol{r}',t)\delta(\boldsymbol{r}'-\boldsymbol{r})d\boldsymbol{r}' = -\psi^*(\boldsymbol{r},t)\Delta_r\psi(\boldsymbol{r},t)$$

$$= \int\nabla_{r'}\psi^*(\boldsymbol{r}',t)\cdot\nabla_{r'}\psi(\boldsymbol{r}',t)\delta(\boldsymbol{r}'-\boldsymbol{r})d\boldsymbol{r}' = \nabla_r\psi^*(\boldsymbol{r},t)\cdot\nabla_r\psi(\boldsymbol{r},t),$$

where the second form follows from a straightforward integration by parts.

A.3. PROBABILITY CONTINUITY

The probability density and current together determine the local balance of the probability distribution in quantum mechanics, as summarized by the probability continuity equation:

$$\frac{\partial p}{\partial t} = -\nabla \cdot \boldsymbol{j} \qquad \text{or}$$

$$\frac{dp}{dt} \equiv \dot{p} = \frac{\partial p}{\partial t} + \nabla \cdot \boldsymbol{j} = \frac{\partial p}{\partial t} + p\nabla \cdot \boldsymbol{V} + \boldsymbol{V} \cdot \nabla p = 0, \qquad \text{(A.21)}$$

which is readily derived as the difference between Eq. (A.12) and its complex conjugate multiplied by ψ^* and ψ, respectively. Above, the total derivative dp/dt is to be interpreted as the time rate of change of the probability density in a volume element of the probability fluid as it moves in space. By contrast, the partial derivative $\partial p/\partial t$ represents the rate of change of the fluid at the fixed point in space. The first form of the preceding equation expresses the fact that for the norm conserving evolution of the system wave function the local rate of change of the probability density is determined solely by the probability density leaving that location, so that the local net production (source) $\dot{p}$ of the probability density identically vanishes, which is explicitly expressed by the second form of this equation.

In the stationary state, representing the eigenfunction ψ of the Hamiltonian (A.4), $\hat{\mathrm{H}}\psi = E\psi$, which corresponds to the sharply-specified energy E, the real amplitude function $\psi = R$ and the phase factor $\Theta = -Et/\hbar \equiv -\omega t$ then directly imply $\boldsymbol{j} = I[\boldsymbol{j}] = 0$. It also follows from the preceding equation that for the complex amplitude of the stationary state, when $\psi \neq \psi^*$, e.g., in the electronically degenerate state, the non-vanishing current $\boldsymbol{j}[\psi] \neq 0$ defines the source-less vector field corresponding to the time-independent probability distribution: $-\partial p/\partial t = \nabla \cdot \boldsymbol{j} = 0$.

An alternative form of the continuity equation follows from expressing the probability current in terms of the probability-fluid velocity of Eq. (A.7), when

$$\frac{dp}{dt} = \frac{\partial p}{\partial t} + \frac{\partial p}{\partial \boldsymbol{r}}\frac{d\boldsymbol{r}}{dt} = \frac{\partial p}{\partial t} + (\nabla p) \cdot \boldsymbol{V}. \qquad \text{(A.22)}$$

Hence, Eq. (A.21) is equivalent to

$$\frac{dp}{dt} + p\nabla \cdot \boldsymbol{V} = 0, \qquad \text{(A.23)}$$

which for the source-less probability fluid, $\dot{p} = 0$, further implies the vanishing divergence of the velocity field: $\nabla \cdot \boldsymbol{V} = 0$. We also observe that the gradient of the probability distribution,

$$\nabla p = (\nabla \psi^*)\psi + \psi^* \nabla \psi = 2\mathrm{Re}(\psi^* \nabla \psi), \qquad \text{(A.24)}$$

measures the *real* (Re) component of the (complex) product $\psi^* \nabla \psi$, the *imaginary* (Im) part of which is proportional to the probability current density of Eq. (A.7).

The quantum mechanical action integral of Eq. (A.13) also becomes the functional of the system time-dependent density, $\mathcal{A} = \mathcal{A}[\rho]$, and the two "hydrodynamical" equations (A.18) and (A.21) are equivalent to the TDDFT analog [81] of the stationary-action principle of Eq. (A.14):

$$\frac{\delta \mathcal{A}[\rho]}{\delta \rho(\boldsymbol{r}, t)} = 0 . \qquad \text{(A.25)}$$

These "hydrodynamical" equations also imply that the time-evolution of the average information content $\bar{I}[\psi] = \bar{I}[p, \boldsymbol{j}]$ of electronic distributions in molecules is effected through the time-dependence of both p and $\boldsymbol{j}$, which together define the independent channels for the information redistribution in molecules. Only at the equilibrium, ground-state the information becomes the unique functional of the ground-state probability distribution [81]:

$$\bar{I}[\psi_{g.s.}] = \bar{I}[p_{g.s.}, \boldsymbol{j}_{g.s.}[p_{g.s.}]] = \bar{I}[p_{g.s.}] .$$

A.4. STATIONARY-STATE RELATIONS

Let us now explore the direct relations linking this information measure with the probability distribution $p(\boldsymbol{r})$, which are implied by the mutually complex-conjugate stationary Schrödinger equations:

$$\hat{\mathrm{H}}(\boldsymbol{r})\psi(\boldsymbol{r}) = E\psi(\boldsymbol{r}) \quad \text{and} \quad \hat{\mathrm{H}}(\boldsymbol{r})\psi^*(\boldsymbol{r}) = E\psi^*(\boldsymbol{r}) . \qquad \text{(A.26)}$$

We first observe that the divergence of Eq. (A.24) gives

$$\Delta p = 2\bar{f} + \psi\Delta\psi^* + \psi^*\Delta\psi. \tag{A.27}$$

In order to eliminate the Laplacian terms in r.h.s. of the preceding equation we use Eqs. (A.26). Multiplying the first equation by ψ^* and the second equation by ψ and taking the sum of the resulting equations gives:

$$\psi\Delta\psi^* + \psi^*\Delta\psi = (4m_e/\hbar^2)(v - E)p, \tag{A.28}$$

and hence

$$\Delta p + (4m_e/\hbar^2)[E - v]p = 2\bar{f}. \tag{A.29}$$

In the spirit of the ordinary DFT) [40-44] this equation expresses the information density in terms of the stationary probability distribution and its Laplacian. The difference between these wave-function-multiplied Eqs. (A.26) similarly gives the familiar relation $\nabla \cdot \boldsymbol{j} = 0$ which confirms the stationary character of the source-less probability distribution.

Alternative relation, between v and $\boldsymbol{j}$, which involves the gradient of the Hamiltonian, follows from the gradient forms of Eqs. (A.26),

$$(\nabla\hat{\mathrm{H}})\psi + \hat{\mathrm{H}}(\nabla\psi) \equiv [\nabla, \hat{\mathrm{H}}]_+\psi = E(\nabla\psi) \quad \text{and}$$

$$(\nabla\hat{\mathrm{H}})\psi^* + \hat{\mathrm{H}}(\nabla\psi^*) \equiv [\nabla, \hat{\mathrm{H}}]_+\psi^* = E(\nabla\psi^*), \tag{A.30}$$

where the anti-commutator

$$[\nabla, \hat{\mathrm{H}}]_+ \equiv \nabla\hat{\mathrm{H}} + \hat{\mathrm{H}}\nabla = -\frac{\hbar^2}{m_e}\nabla^3 + 2v\nabla + \nabla v = 2\hat{\mathrm{H}}\nabla + \nabla v, \tag{A.31}$$

and ∇v denotes the multiplication by the gradient of the external potential. Taking the difference of products of both sides of the first equation with $\nabla\psi^*$ and of the second equation with $\nabla\psi$, respectively, then gives the following stationary Schrödinger relation:

$$(\nabla v) \cdot \boldsymbol{j} = 0, \tag{A.32}$$

where we have recognized the locally Hermitian character of the Hamiltonian:

$$\int \nabla_{r'}\psi^*(\boldsymbol{r}',t)\cdot \hat{\mathrm{H}}(\boldsymbol{r}')\nabla_{r'}\psi(\boldsymbol{r}',t)\delta(\boldsymbol{r}'-\boldsymbol{r})d\boldsymbol{r}'$$

$$= \int \nabla_{r'}\psi(\boldsymbol{r}',t)\cdot \hat{\mathrm{H}}(\boldsymbol{r}')\nabla_{r'}\psi^*(\boldsymbol{r}',t)\delta(\boldsymbol{r}'-\boldsymbol{r})d\boldsymbol{r}'$$

$$= \nabla_r\psi(\boldsymbol{r},t)\cdot \hat{\mathrm{H}}(\boldsymbol{r})\nabla_r\psi^*(\boldsymbol{r},t). \qquad \text{(A.33)}$$

Equation (A.32) indicates that in the stationary state the directions of $\nabla v = -\boldsymbol{F}$ (external force acting on electron) and $\boldsymbol{j}$ are mutually perpendicular.

For the real (electronically non-degenerate) stationary states, when $p(\boldsymbol{r})$ and $\bar{f}(\boldsymbol{r})$ are stationary and $\boldsymbol{j}$ identically vanishes, the amount of information is conserved in time. It also follows from the definition of the information density $\bar{f} = \nabla\psi\cdot\nabla\psi^*$ and the associated differential,

$$\delta\bar{f} = \frac{\partial\bar{f}}{\partial\psi}\delta\psi + \frac{\partial\bar{f}}{\partial\psi^*}\delta\psi^* = -(\Delta\psi^*)\delta\psi - (\Delta\psi)\delta\psi^*, \qquad \text{(A.34)}$$

that the Laplacian terms of Eq. (A.28) can be expressed in terms of the partial derivatives of $\bar{f}$ with respect to ψ and ψ^*, respectively, thus giving the alternative form of Eq. (A.27):

$$2\bar{f} - \left(\frac{\partial\bar{f}}{\partial\psi}\psi + \frac{\partial\bar{f}}{\partial\psi^*}\psi^*\right) = \Delta p. \qquad \text{(A.35)}$$

This differential equation links the information density to its linear homogeneity with respect to the wave-function arguments (ψ, ψ^*) and the probability distribution p.

A.5. INFORMATION BALANCE

For an interpretation of the information transport effected by the electron redistribution in molecules the concepts of the information flux and source are paramount [24]. It should be stressed that the overall amount of information, proportional to the system average kinetic energy, is not conserved in a general *non*-stationary state, when the probability density evolves in time. Indeed, different wave-functions give rise to different expectation values of the kinetic energy. Therefore, the continuity equation expressing the local balance of the Fisher information in quantum mechanics has to include a *non*-vanishing "source" term $df/dt = \dot{f} \neq 0$ [24]:

$$\frac{\partial \bar{f}}{\partial t} = \dot{\bar{f}} - \nabla \cdot \bar{\boldsymbol{J}} \quad \text{or} \quad \frac{d\bar{f}}{dt} \equiv \dot{\bar{f}} = \frac{\partial \bar{f}}{\partial t} + \nabla \cdot \bar{\boldsymbol{J}} \neq 0 . \tag{A.36}$$

Above, the renormalized information-current density $\bar{\boldsymbol{J}}[\psi]$ for the specified quantum state ψ and the associated source is to be identified using the known dynamics of quantum states determined by the time-dependent Schrödinger equation (A.12) and its complex conjugate. It can be again physically interpreted as the product of the information density $\bar{f}[\psi]$ and the "fluid" local velocity $\boldsymbol{V}$:

$$\bar{\boldsymbol{J}}[\psi] = \bar{f}[\psi]\, \boldsymbol{V}[\psi] = \left(\frac{\bar{f}}{p}\right) \boldsymbol{j}[\psi]. \tag{A.37}$$

It follows from this expression that

$$\nabla \cdot \bar{\boldsymbol{J}} = (\nabla \bar{f}) \cdot \boldsymbol{V} + \bar{f}\ \nabla \cdot \boldsymbol{V} = (\nabla \bar{f}) \cdot \boldsymbol{V},$$

$$(\nabla \bar{f}) = (\Delta \psi^*) \nabla \psi + (\nabla \psi^*) \Delta \psi, \tag{A.38}$$

where we have used the implication of Eq. (A.23) that $\nabla \cdot \boldsymbol{V} = 0$. Therefore, the source $\dot{\bar{f}}$ of the quantum Fisher information can be expressed as

$$\frac{d\bar{f}}{dt} = \frac{\partial \bar{f}}{\partial t} + (\nabla \bar{f}) \cdot \frac{\boldsymbol{j}}{p}. \tag{A.39}$$

The partial derivative term can be determined using the time-dependent Schrödinger equation (A.12) and its complex conjugate, which together imply the partial derivative expression of Eq. (A.17). For $\hat{Q} = \hat{\bar{f}}(\boldsymbol{r})$ the latter directly implies:

$$\frac{\partial \bar{f}(\boldsymbol{r})}{\partial t} = \frac{i}{\hbar}\langle\psi(t)|[\hat{H}, \hat{\bar{f}}(\boldsymbol{r})]|\psi(t)\rangle = \frac{i}{\hbar}\langle\psi(t)|[\Delta, \hat{H}]|\psi(t)\rangle$$

$$= \frac{i}{\hbar}\langle\psi(t)|[\Delta, v]|\psi(t)\rangle, \tag{A.40}$$

where the commutator $[\Delta, v] = 2(\nabla v)\cdot\nabla + (\Delta v)$. Hence,

$$\frac{\partial \bar{f}(\boldsymbol{r},t)}{\partial t} = \frac{i}{\hbar}\{[\nabla v(\boldsymbol{r})]\cdot 2\psi^*(\boldsymbol{r},t)\nabla\psi(\boldsymbol{r},t) + [\Delta v(\boldsymbol{r})]\psi^*(\boldsymbol{r},t)\psi(\boldsymbol{r},t)\}$$

$$= \frac{i}{\hbar}\{[\nabla v(\boldsymbol{r})]\cdot 2\psi^*(\boldsymbol{r},t)\nabla\psi(\boldsymbol{r},t) + [\Delta v(\boldsymbol{r})]\,p(\boldsymbol{r},t)\}. \tag{A.41}$$

In order to identify the physical meaning of the first term in the preceding equation one uses Eqs. (A.7) and (A.24) which give:

$$2\psi^*\nabla\psi = \frac{2m_e i}{\hbar}\boldsymbol{j} + \nabla p. \tag{A.42}$$

Inserting this expression into Eq. (A.41) then implies the following physical interpretation of the latter:

$$\frac{\partial \bar{f}}{\partial t} = \frac{i}{\hbar}\left(\frac{2m_e i}{\hbar}\boldsymbol{j}\cdot\nabla v + \nabla p\cdot\nabla v + p\Delta v\right) = -\frac{2m_e}{\hbar^2}\nabla v\cdot\boldsymbol{j}. \tag{A.43}$$

Above we have used the local identity which follows from the integration by parts:

$$\nabla p(\boldsymbol{r},t)\cdot\nabla v(\boldsymbol{r}) = \int \nabla_{r'} p(\boldsymbol{r}',t)\cdot\nabla_{r'} v(\boldsymbol{r}')\delta(\boldsymbol{r}'-\boldsymbol{r})d\boldsymbol{r}'$$

$$= -\int p(\boldsymbol{r}',t)\Delta_{r'} v(\boldsymbol{r}')\delta(\boldsymbol{r}'-\boldsymbol{r})d\boldsymbol{r}' = -p(\boldsymbol{r},t)\Delta_r v(\boldsymbol{r}).$$

Equation (A.43) can be alternatively derived from the gradient forms of the time-dependent Schrödinger equation and its complex conjugate, respectively [compare Eqs. (A.30]:

$$[\nabla, \hat{\mathrm{H}}]_+ \psi = i\hbar \frac{\partial \nabla \psi}{\partial t} \qquad \text{and} \qquad [\nabla, \hat{\mathrm{H}}]_+ \psi^* = -i\hbar \frac{\partial \nabla \psi^*}{\partial t}, \tag{A.44}$$

where the anti-commutator has been defined in Eq. (A.31). Multiplying both sides of these two equations by $\nabla \psi^*$ and $\nabla \psi$, respectively, and subtracting the resulting equations gives:

$$i\hbar\left(\nabla \psi^* \cdot \frac{\partial \nabla \psi}{\partial t} + \frac{\partial \nabla \psi^*}{\partial t} \cdot \nabla \psi\right) = i\hbar \frac{\partial \bar{f}}{\partial t}$$

$$= (\nabla \psi^*) \cdot [2\hat{\mathrm{H}}(\nabla \psi) + \psi \nabla v] - (\nabla \psi) \cdot [2\hat{\mathrm{H}}(\nabla \psi^*) + \psi^* \nabla v]$$

$$= \frac{2m_e}{\hbar^2} \nabla v \cdot \boldsymbol{j}\,. \tag{A.45}$$

where we have used the Hermitian property of the Hamiltonian [Eq. (A.33)].

Finally, substituting Eq. (A.43) into Eq. (A.39) gives the following expression for the Fisher information source:

$$\frac{d\bar{f}}{dt} = \frac{\partial \bar{f}}{\partial t} + \frac{\nabla \bar{f}}{p} \cdot \boldsymbol{j} = \left(\frac{2m_e}{\hbar^2} \nabla v + \frac{\nabla \bar{f}}{p}\right) \cdot \boldsymbol{j} \equiv \boldsymbol{F}_f \cdot \boldsymbol{j}\,. \tag{A.46}$$

It is proportional to the probability current density, locally weighted by the sum of contributions due to ∇v, the negative external force acting on electron, and the gradient of information density per electron, $p^{-1}\nabla f$. In the spirit of the ordinary irreversible thermodynamics [82] it is given by the product of the particle flux $\boldsymbol{j}$ and the vector information "force" $\boldsymbol{F}_f$ which combines the external force acting on electron, $\boldsymbol{F} = -\nabla v$, and the gradient of the information density itself, $\nabla \bar{f}$. The continuity equation for the Fisher information determines the dynamics of the information density in quantum systems thus offering an altogether new perspective on the information redistributions accompanying chemical changes in molecular world.

Appendix B

CONDITIONAL PROBABILITIES FROM THE BOND-PROJECTED SUPERPOSITION PRINCIPLE

The conditional probabilities of quantum states are embodied in the *superposition principle* of quantum mechanics [45]. They involve squares of the moduli of the relevant basis-set expansion coefficients, which represent projections of one state onto another. For example, the probability $P(\psi|\phi)$ of the *variable* state ψ conditional on another *parameter* state ϕ is given by the square of the corresponding overlap integral $\langle\phi|\psi\rangle$,

$$P(\psi|\phi) = |\langle\phi|\psi\rangle|^2 = \langle\psi|\phi\rangle\langle\phi|\psi\rangle. \tag{B.1}$$

It can be expressed as the expectation value of the state projection operators:

$$P(\psi|\phi) = \langle\psi|\phi\rangle\langle\phi|\psi\rangle \equiv \langle\psi|\hat{P}_\phi|\psi\rangle = \langle\phi|\psi\rangle\langle\psi|\phi\rangle \equiv \langle\phi|\hat{P}_\psi|\phi\rangle = P(\phi|\psi), \tag{B.2}$$

where the conditional probability of the state upon itself $P(\psi|\psi) = P(\phi|\phi) = 1$, since for normalized states $\langle\psi|\psi\rangle = \langle\phi|\phi\rangle = 1$. This *two*-state conditional probability is thus determined by the expectation value in the *variable* state, say ψ, of the projection operator $\hat{P}_\phi$ onto the *parameter* (reference) state ϕ.

The conditional probability concept constitutes the basis of the superposition principle, that any combination $|\Psi\rangle = \sum_i C_i |\Psi_i\rangle$ of the complete basis of quantum states $\mathbf{\Psi} = \{|\Psi_i\rangle\}$ itself represents a possible quantum state of the system under consideration. The squares of the moduli of the expansion coefficients $\{C_i = \langle\Psi_i|\Psi\rangle\}$, in general complex numbers, determine the conditional probabilities $\{P(\Psi_i|\Psi) = |C_i|^2\}$ of observing state $|\Psi_i\rangle$ given the state $|\Psi\rangle$, thus satisfying the normalization condition

$$\sum_i P(\Psi_i|\Psi) = \langle\Psi|(\sum_i |\Psi_i\rangle\langle\Psi_i|)|\Psi\rangle = \sum_i |C_i|^2 = \langle\Psi|\Psi\rangle = 1, \quad (B.3)$$

since the projection of the state vector $|\Psi\rangle$ onto the Hilbert space spanned by the complete basis amounts to identity operation:

$$(\sum_i |\Psi_i\rangle\langle\Psi_i|)\,|\Psi\rangle \equiv \hat{P}_\Psi|\Psi\rangle = |\Psi\rangle. \quad (B.4)$$

In OCT we are interested in the mutual dependencies between AO, which originate from the orbital involvement in the whole system of chemical bonds in the molecule determined by the *occupied* (o) MO $\boldsymbol{\varphi}$, a subspace in the complete set of MO derived from the adopted basis functions $\boldsymbol{\chi}$, $\boldsymbol{\varphi}^{o+v} = (\boldsymbol{\varphi}, \boldsymbol{\varphi}^v)$, where $\boldsymbol{\varphi}^v$ determines the complementary (virtual MO) subspace:

$$\hat{P}_\chi \equiv |\chi\rangle\langle\chi| = |\varphi^{o+v}\rangle\langle\varphi^{o+v}| \equiv \hat{P}_\varphi^{o+v} = |\varphi\rangle\langle\varphi| + |\varphi^v\rangle\langle\varphi^v| = \hat{P}_\varphi + \hat{P}_\varphi^v. \quad (B.5)$$

These *physical* conditional probabilities of AO can be thus thought of as involving additional projection $\hat{P}_\varphi$ onto the molecular Hilbert space spanned by $\boldsymbol{\varphi}$,

$$P(j|i) = 4\mathcal{N}_i\,|\langle i|\hat{P}_\varphi|j\rangle|^2 = 4\mathcal{N}_i\langle i|\hat{P}_\varphi|j\rangle\langle j|\hat{P}_\varphi|i\rangle = (2\gamma_{i,i})^{-1}\gamma_{i,j}\gamma_{j,i},$$
$$(i, j) = 1, 2, ..., m, \quad (B.6)$$

where the closed-shell normalization constant $\mathcal{N}_i = (2\gamma_{i,i})^{-1}$ follows directly from the idempotency relation of the CBO matrix Eq. (199), involving the double occupancy of all states $\boldsymbol{\varphi}$. In OCT the first (variable) index j in $P(j|i)$ identifies the "output" AO event, while the second (parameter) index i specifies the given "input" AO event; these indices determines the columns and rows in the associated probability matrix $\mathbf{P}(\chi|\chi) = \{P(j|i)\}$, which defines the AO communication channel. These conditional probabilities determine the probability scattering in the AO-promotion channel of the molecule, in which the "signals" of the molecular (or promolecular) electron allocations to basis functions are transmitted between the AO inputs $\boldsymbol{a}$ and outputs $\boldsymbol{b}$.

The *bond*-conditional probabilities $\mathbf{P}(\chi|\chi) \equiv \mathbf{P}(\boldsymbol{b}|\boldsymbol{a})$ subsequently define the associated joint probabilities of the simultaneous *two*-AO probabilities $\mathbf{P}(\chi\wedge\chi) = \{P(i\wedge j)\} \equiv \mathbf{P}(\boldsymbol{a}\wedge\boldsymbol{b})$, for the specified pair of the input and output AO events,

$$P(i\wedge j) = p_i P(j|i) = (2N)^{-1}\gamma_{i,j}\gamma_{j,i}, \qquad (i, j) = 1, 2, ..., m. \tag{B.7}$$

As we have already emphasized above, the normalization constant $\mathcal{N}_i = (2\gamma_{i,i})^{-1}$ of Eq. (B.6) applies only to the molecular closed-shell configurations. Indeed, by using the idempotency relation of Eq. (199), one then directly verifies the relevant normalization condition:

$$\sum_j P(j|i) = (2\gamma_{i,i})^{-1}\sum_j \gamma_{i,j}\gamma_{j,i} = 1. \tag{B.8}$$

In the (single-determinant) open-shell case one partitions the CBO (density) matrix into contributions originating from the closed-shell (doubly-occupied) MO φ_β and the open-shell (singly-occupied) MO φ_α, $\varphi = (\varphi_\alpha, \varphi_\beta)$:

$$\gamma = \langle\chi|\varphi_\alpha\rangle\langle\varphi_\alpha|\chi\rangle + 2\langle\chi|\varphi_\beta\rangle\langle\varphi_\beta|\chi\rangle \equiv \langle\chi|\hat{\mathrm{P}}_\varphi^\alpha|\chi\rangle + 2\langle\chi|\hat{\mathrm{P}}_\varphi^\beta|\chi\rangle \equiv \gamma^\alpha + \gamma^\beta. \tag{B.9}$$

They satisfy separate idempotency relations,

$$(\gamma^\alpha)^2 = \langle\chi|\hat{\mathrm{P}}_\varphi^\alpha|\chi\rangle\langle\chi|\hat{\mathrm{P}}_\varphi^\alpha|\chi\rangle = \langle\chi|(\hat{\mathrm{P}}_\varphi^\alpha)^2|\chi\rangle = \langle\chi|\hat{\mathrm{P}}_\varphi^\alpha|\chi\rangle = \gamma^\alpha,$$

$$(\gamma^\beta)^2 = 4\langle\chi|\hat{\mathrm{P}}_\varphi^\beta|\chi\rangle\langle\chi|\hat{\mathrm{P}}_\varphi^\beta|\chi\rangle = 4\langle\chi|(\hat{\mathrm{P}}_\varphi^\beta)^2|\chi\rangle = 4\langle\chi|\hat{\mathrm{P}}_\varphi^\beta|\chi\rangle = 2\gamma^\beta, \tag{B.10}$$

where we have recognized the idempotency/orthogonality of the MO projections, $\hat{\mathrm{P}}_\varphi^\zeta\hat{\mathrm{P}}_\varphi^\xi = \hat{\mathrm{P}}_\varphi^\xi\delta_{\zeta,\xi}$, and the identity character of the overall AO projection, $\hat{\mathrm{P}}_\chi \equiv |\chi\rangle\langle\chi| = 1$, when acting on MO.

Hence, one finds the generalized normalization constant $\mathcal{N}_i = (\gamma_{i,i}^\alpha + 2\gamma_{i,i}^\beta)^{-1}$ in the conditional probability

$$P(j|i) = (\gamma_{i,i}^\alpha + 2\gamma_{i,i}^\beta)^{-1}\gamma_{i,j}\gamma_{j,i}, \tag{B.11}$$

which satisfies the sum-rule for conditional probabilities in *i*th row of $\mathbf{P}(\chi|\chi)$:

$$\sum_j P(j|i) = (\gamma_{i,i}^{\alpha} + 2\gamma_{i,i}^{\beta})^{-1} \sum_j (\gamma_{i,j}^{\alpha} + \gamma_{i,j}^{\beta})(\gamma_{j,i}^{\alpha} + \gamma_{j,i}^{\beta})$$

$$= (\gamma_{i,i}^{\alpha} + 2\gamma_{i,i}^{\beta})^{-1} \sum_j (\gamma_{i,j}^{\alpha}\gamma_{j,i}^{\alpha} + \gamma_{i,j}^{\beta}\gamma_{j,i}^{\beta})$$

$$= (\gamma_{i,i}^{\alpha} + 2\gamma_{i,i}^{\beta})^{-1} (\gamma_{i,i}^{\alpha} + 2\gamma_{i,i}^{\beta}) = 1. \tag{B.12}$$

Above, we have used the idempotency relations of Eq. (A.10) and realized that

$$\sum_j \gamma_{i,j}^{\alpha}\gamma_{j,i}^{\beta} = 2\langle i|\hat{\mathrm{P}}_{\varphi}^{\alpha}|\chi\rangle \langle \chi|\hat{\mathrm{P}}_{\varphi}^{\beta}|i\rangle = \langle i|\hat{\mathrm{P}}_{\varphi}^{\alpha}\hat{\mathrm{P}}_{\varphi}^{\beta}|i\rangle$$

$$= \sum_j \gamma_{i,j}^{\beta}\gamma_{j,i}^{\alpha} = \langle i|\hat{\mathrm{P}}_{\varphi}^{\beta}\hat{\mathrm{P}}_{\varphi}^{\alpha}|i\rangle = 0. \tag{B.13}$$

This open-shell development can be straightforwardly generalized into the case of the (equal) fractional occupations of the open-shell MO, which result from the ensemble averaging of the above CBO contributions due to integer MO occupations.

Appendix C

MANY–STATE SUPERPOSITION PERSPECTIVE

As we have shown in the preceding appendix the concept of the *two*-state conditional probability $P(\phi|\psi) = \langle\psi|\hat{P}_\phi|\psi\rangle \equiv |\psi_\phi|^2$ emerges in the context of an expansion of one state in the complete basis set $\{|\phi\rangle\}$ containing the other state,

$$|\psi\rangle = \Sigma_\phi |\phi\rangle\langle\phi|\psi\rangle = \Sigma_\phi |\phi\rangle \psi_\phi, \qquad \text{(C.1)}$$

where the (complex) expansion coefficient ψ_ϕ reflects the projection of $|\psi\rangle$ on $|\phi\rangle$ in the vector (Hilbert) space of molecular states. In the physical scenario of the molecular bond system, defined by the occupied MO $|\varphi\rangle$, the additional projection $\hat{P}_\varphi$ onto this subspace intervenes in the expression for the bond-conditional probability between two AO [Eq. (B.6)]:

$$P(j|i) = \mathcal{N}_i\, |\langle i|\hat{P}_\varphi|j\rangle|^2 = \mathcal{N}_i \langle i|\hat{P}_\varphi|j\rangle\langle j|\hat{P}_\varphi|i\rangle = \mathcal{N}_i \langle i|\hat{P}_\varphi|j\rangle\langle j|\hat{P}_\varphi|i\rangle \,. \qquad \text{(C.2)}$$

The *doubly*-conditional relations, say between ψ and χ through the basis functions $\{\phi\}$, involve three quantum states. They appear when one expands the basis vector $|\phi\rangle$ in the preceding equation in terms of another basis set $\{|\chi\rangle\}$:

$$|\psi\rangle = \Sigma_\chi |\chi\rangle\, (\Sigma_\phi \langle\chi|\phi\rangle\langle\phi|\psi\rangle) \equiv \Sigma_\chi |\chi\rangle\, (\Sigma_\phi\, \psi_{\chi,\phi}) = \Sigma_\chi |\chi\rangle \psi_\chi, \qquad \text{(C.3)}$$

where:

$$P(\chi|\psi) = \langle\psi|\hat{\mathrm{P}}_\chi|\psi\rangle = |\psi_\chi|^2 = |\Sigma_\phi\, \psi_{\chi,\phi}|^2 \equiv P(\{\phi\}||\chi|\psi)$$

$$= |\Sigma_\phi\langle\chi|\phi\rangle\langle\phi|\psi\rangle|^2 = |\Sigma_\phi\, \phi_\chi\psi_\phi|^2$$

$$= \Sigma_{\phi'}\Sigma_\phi\langle\psi|\phi'\rangle\langle\phi'|\chi\rangle\langle\chi|\phi\rangle\langle\phi|\psi\rangle = \Sigma_{\phi'}\Sigma_\phi\langle\psi|\hat{\mathrm{P}}_{\phi'}\hat{\mathrm{P}}_\chi\hat{\mathrm{P}}_\phi|\psi\rangle$$

$$= \sum_{\phi'}\sum_{\phi}\phi'_\psi\,\chi_{\phi'}\phi_\chi\psi_\phi\,. \tag{C.4}$$

In the physical context of the mutual couplings between three AO in the molecular bond system the additional projections into the occupied MO subspace are called for:

$$P(\chi|\psi||\{\phi,\phi'\}) = \mathcal{M}_\psi\, \Sigma_\phi\Sigma_{\phi'}\langle\psi|\hat{\mathrm{P}}_\varphi|\phi'\rangle\langle\phi'|\hat{\mathrm{P}}_\varphi|\chi\rangle\langle\chi|\hat{\mathrm{P}}_\varphi|\phi\rangle\langle\phi|\hat{\mathrm{P}}_\varphi|\psi\rangle$$

$$= \mathcal{M}_\psi\, \Sigma_\phi\Sigma_{\phi'}\gamma_{\psi,\phi'}\gamma_{\phi',\chi}\gamma_{\chi,\phi}\gamma_{\phi,\psi}, \tag{C.5}$$

where $\mathcal{M}_\psi$ stands for the normalization constant satisfying the condition of Eq. (C.4) [see also Eqs (199) and (B.6)]:

$$\mathcal{M}_\psi\, \Sigma_\phi\Sigma_{\phi'}\gamma_{\psi,\phi'}\gamma_{\phi',\chi}\gamma_{\chi,\phi}\gamma_{\phi,\psi} = 4\,\mathcal{M}_\psi\gamma_{\psi,\chi}\gamma_{\chi,\psi} = P(\chi|\psi) = (2\gamma_{\psi,\psi})^{-1}\gamma_{\psi,\chi}\gamma_{\chi,\psi}, \tag{C.6}$$

or $\mathcal{M}_\psi = (8\gamma_{\psi,\psi})^{-1}$.

The individual terms in the above double summation, which can assume negative values, define the *double*-conditional "probability" matrix

$$\mathbf{P}(\chi|\psi||\mathbf{\Phi}) = \{P(\chi|\psi||\{\phi,\phi'\})\}. \tag{C.7}$$

Its diagonal element,

$$P(\chi|\psi||\phi,\phi) \equiv P(\chi|\psi||\phi) = \mathcal{M}_\psi\,(\gamma_{\phi,\psi}\gamma_{\psi,\phi})(\gamma_{\phi,\chi}\gamma_{\chi,\phi})$$

$$= (8\gamma_{\psi,\psi})^{-1}|\gamma_{\phi,\psi}|^2|\gamma_{\chi,\phi}|^2 \ge 0, \tag{C.8}$$

can thus be regarded as an admissible probability $P(\chi|\psi||\phi)$, of the $\psi\rightarrow\chi$ communication in the system chemical bonds, conditional on ϕ.

Indeed, the individual (complex) expansion coefficient in Eq. (C.3),

$$\psi_{\chi,\phi} = \langle\chi|\phi\rangle\langle\phi|\psi\rangle = \phi_\chi\psi_\phi, \tag{C.9}$$

can itself be regarded as the amplitude of the elementary *doubly*-conditional probability, of χ on ψ though ϕ,

$$\pi(\chi|\psi||\phi) \equiv |\psi_{\chi,\phi}|^2 = \langle\psi|\phi\rangle\langle\phi|\chi\rangle\langle\chi|\phi\rangle\langle\phi|\psi\rangle = \langle\psi|\hat{P}_\phi\hat{P}_\chi\hat{P}_\phi|\psi\rangle. \tag{C.10}$$

Its bond-projected analog again calls for the additional projections onto the occupied MO subspace:

$$\pi(\chi|\psi||\phi) = \mathcal{M}_\psi\langle\psi|\hat{P}_\varphi|\phi\rangle\langle\phi|\hat{P}_\varphi|\chi\rangle\langle\chi|\hat{P}_\varphi|\phi\rangle\langle\phi|\hat{P}_\varphi|\psi\rangle = P(\chi|\psi||\phi). \tag{C.11}$$

Therefore, these double-conditional probabilities between three AO in the molecular bond system is given by the product of squares of CBO matrix elements coupling the two AO involved in the elementary communication ψ(*parameter*)$\to\chi$(*variable*) with the third orbital ϕ (*intermediate*), divided by diagonal matrix element (population) of the parameter AO.

The superposition principle introducing the conditional dependencies between four quantum states, calls for expanding $|\chi\rangle$ in Eq. (C.3) in terms of basis vectors $\{|\xi\rangle\}$

$$|\psi\rangle = \Sigma_\xi|\xi\rangle\,(\Sigma_\chi\,\Sigma_\phi\langle\xi|\chi\rangle\langle\chi|\phi\rangle\langle\phi|\psi\rangle) \equiv \Sigma_\xi|\xi\rangle\,(\Sigma_\chi\,\Sigma_\phi\,\psi_{\chi,\phi,\chi}) = \Sigma_\chi|\xi\rangle\psi_\xi\,. \tag{C.12}$$

Appropriately normalized (bond-projected) squares of the individual expansion coefficients $\psi_{\chi,\phi,\chi}$ can now be regarded as measuring the triply conditional probability, of the dependence of the ψ(*parameter*) $\to$ ξ(*variable*) communication on the two intermediate states (ϕ and χ):

$$\mathcal{L}_\psi\langle\psi|\hat{P}_\varphi|\phi\rangle\langle\phi|\hat{P}_\varphi|\chi\rangle\langle\chi|\hat{P}_\varphi|\xi\rangle\langle\xi|\hat{P}_\varphi|\chi\rangle\langle\chi|\hat{P}_\varphi|\phi\rangle\langle\phi|\hat{P}_\varphi|\psi\rangle \equiv P(\xi|\psi||\phi,\chi)$$

$$= \mathcal{L}_\psi\langle\psi|\hat{P}_\varphi\,\hat{P}_\phi\,\hat{P}_\varphi\,\hat{P}_\chi\,\hat{P}_\varphi\,\hat{P}_\xi\,\hat{P}_\varphi\,\hat{P}_\chi\,\hat{P}_\varphi\,\hat{P}_\phi\,\hat{P}_\varphi|\psi\rangle$$

$$= \mathcal{L}_\psi(\gamma_{\phi,\psi}\gamma_{\psi,\phi})(\gamma_{\phi,\chi}\gamma_{\chi,\phi})(\gamma_{\xi,\chi}\gamma_{\chi,\xi}) = \mathcal{L}_\psi|\gamma_{\phi,\psi}|^2\,|\gamma_{\chi,\phi}|^2\,|\gamma_{\xi,\chi}|^2 \geq 0. \tag{C.13}$$

The normalization constant $\mathcal{L}_\psi$ is chosen to recover $P(\xi|\psi) = |\psi_\xi|^2$:

$$\mathcal{L}_\psi\Sigma_{\phi'}\Sigma_\phi\,\gamma_{\psi,\phi'}\gamma_{\phi,\psi}\,(\Sigma_{\chi'}\,\gamma_{\phi',\chi'}\,\gamma_{\chi',\xi})\,(\Sigma_\chi\,\gamma_{\xi,\chi}\gamma_{\chi,\phi})$$

$$= 4\mathcal{L}_\psi(\Sigma_{\phi'}\,\gamma_{\psi,\phi'}\,\gamma_{\phi',\xi})(\Sigma_\phi\,\gamma_{\xi,\phi}\gamma_{\phi,\psi})$$

$$= 16\mathcal{L}_{\psi}\gamma_{\psi,\xi}\gamma_{\xi,\psi} = P(\xi|\psi) = (2\gamma_{\psi,\psi})^{-1}\gamma_{\psi,\xi}\gamma_{\xi,\psi}, \tag{C.14}$$

and hence $\mathcal{L}_{\psi} = (32\gamma_{\psi,\psi})^{-1}$.

To summarize, it follows from Eqs. (C.6) and (C.14) that – to a constant factor – the corresponding powers of the CBO matrix determine the *transition*-amplitudes for the conditional probabilities of the *multi*-orbital events in the molecular bond system, which involve the relevant intermeadiate states, in accordance with the idempotency relations of Eq. (199), e.g.,

$$(\gamma^2)_{\chi,\psi} = 2\gamma_{\chi,\psi}, \quad (\gamma^3)_{\xi,\psi} = \sum_{\chi}\gamma_{\xi,\chi}(\gamma^2)_{\chi,\psi}, \text{ etc.} \tag{C.15}$$

This gives rise to the sequential cascade interpretation of the *many*-orbital information system as the product of intermediate amplitude propagation stages shown in Scheme B.1, with the communication network at each stage determined by the AO density matrix γ.

$$\to\left\{\psi \xrightarrow{\gamma} \left(\varphi' \xrightarrow{\gamma} \chi \xrightarrow{\gamma} \varphi\right) \xrightarrow{\gamma} \psi\right\}\to$$

$$\equiv \to\left\{\psi \xrightarrow{\gamma} \left(\varphi' \xrightarrow{\gamma} \varphi\right) \xrightarrow{\gamma} \psi\right\}\to$$

$$\to\left\{\psi \xrightarrow{\gamma} \left(\varphi' \xrightarrow{\gamma} [\chi' \xrightarrow{\gamma} \xi \xrightarrow{\gamma} \chi] \xrightarrow{\gamma} \varphi\right) \xrightarrow{\gamma} \psi\right\}\to$$

$$\equiv \to\left\{\psi \xrightarrow{\gamma} \left(\varphi' \xrightarrow{\gamma} [\chi' \xrightarrow{\gamma} \chi] \xrightarrow{\gamma} \varphi\right) \xrightarrow{\gamma} \psi\right\}\to$$

$$\equiv \to\left\{\psi \xrightarrow{\gamma} \left(\varphi' \xrightarrow{\gamma} \varphi\right) \xrightarrow{\gamma} \psi\right\}\to$$

Scheme C.1. Cascade interpretation of multi-orbital conditional probabilities from the superposition principle involving one (upper diagrams) and two (lower diagrams) intermediate bases.

Clearly, this procedure can be straightforwardly generalized to any required number of intermediate states.

Appendix D

COUPLING APPROACH TO MANY–ORBITAL BOND INTERACTIONS

To determine the entropy/information quantities of the three dependent probability distributions, which are schematically shown in the diagrams of the lower part of Scheme 4 (Section 1.5), some adequate measures of the *joint* probabilities $\mathbf{P}(\boldsymbol{a}\wedge\boldsymbol{b}\wedge\boldsymbol{c}) = \{P(i\wedge j\wedge k)\}$, of the simultaneous occurrence of the *three*-orbital events in the molecular chemical bond system, are required [20,22]. They must satisfy the relevant partial sum rules,

$$\sum_i P(i\wedge j\wedge k) = P(j\wedge k),\ \sum_j P(i\wedge j\wedge k) = P(i\wedge k),\ \sum_k P(i\wedge j\wedge k) = P(i\wedge j), \tag{D.1}$$

and define the associated *conditional* probabilities through equalities:

$$\mathbf{P}(\boldsymbol{a}\wedge\boldsymbol{b}\wedge\boldsymbol{c}) = \{P(i\wedge j\wedge k) = p_k\, P(i\wedge j|k) = p_j\, P(i\wedge k|j) = p_i\, P(j\wedge k|i)\},$$

$$\sum_i\sum_j P(i\wedge j|k) = \sum_i\sum_k P(i\wedge k|j) = \sum_j\sum_k P(j\wedge k|i) = 1, \tag{D.2}$$

or

$$\mathbf{P}(\boldsymbol{a}\wedge\boldsymbol{b}\wedge\boldsymbol{c}) = \{P(i\wedge j\wedge k) = P(i\wedge j)\, P(k|i\wedge j)$$

$$= P(i\wedge k)\, P(j\,|i\wedge k) = P(j\wedge k)P(i\,|j\wedge k)\},$$

$$\sum_i P(i|\,j\wedge k) = \sum_j P(j\,|i\wedge k) = \sum_k P(k|i\wedge j) = 1. \tag{D.3}$$

Similarly, the entropic descriptors of the four dependent probability distributions in Scheme 5, $\mathbf{P}(\boldsymbol{a})$, $\mathbf{P}(\boldsymbol{b})$, $\mathbf{P}(\boldsymbol{c})$ and $\mathbf{P}(\boldsymbol{d})$, require the *four*-AO probabilities $\mathbf{P}(\boldsymbol{a}\wedge\boldsymbol{b}\wedge\boldsymbol{c}\wedge\boldsymbol{d}) = \{P(i\wedge j\wedge k\wedge l)\}$ [20,22], which in turn define alternative sets of conditional probability propagations describing the (external) bond-couplings between the (internal) AO communications in molecular subsystems [22].

A straightforward extension of Eqs. (204) and (B.7) gives the following expression for the projected measure of the simultaneous coupling of *three* AO in the molecular framework of chemical bonds [20,22]:

$$\mathbf{Q}(\boldsymbol{a}\wedge\boldsymbol{b}\wedge\boldsymbol{c}) = \{Q(i\wedge j\wedge k) = (4N)^{-1}\gamma_{i,j}\gamma_{j,k}\gamma_{k,i} = (4N)^{-1}\langle i|\hat{\mathrm{P}}_{\varphi}|j\rangle\langle j|\hat{\mathrm{P}}_{\varphi}|k\rangle\langle k|\hat{\mathrm{P}}_{\varphi}|i\rangle\},\tag{D.4}$$

and hence the associated conditional overlaps [compare Eqs. (202) and (B.11)]:

$$\mathbf{Q}(\boldsymbol{a}\wedge\boldsymbol{b}|\boldsymbol{c}) = \{Q(i\wedge j|k) = Q(i\wedge j\wedge k)/p_k = (4\gamma_{k,k})^{-1}\gamma_{k,i}\gamma_{i,j}\gamma_{j,k}\}.\tag{D.5}$$

It should be realized, that the electron occupying a given MO $\varphi_s = \boldsymbol{\chi}\boldsymbol{C}_s$, where $\boldsymbol{C}_s = \{C_{i,s}\}$ stands for the sth column in $\mathbf{C} = \{\boldsymbol{C}_s\}$, is simultaneously observed in all AO which participate in this linear combination, i.e., on all such basis functions. The resultant effect from all occupied MO, which determine the network of chemical bonds in the molecule, on such simultaneous probabilities of several AO are obtained by applying the projector $\hat{\mathrm{P}}_{\varphi}$ onto the occupied subspace of MO, which then links such (physical) projected coupling indices to the relevant product of the CBO matrix elements (see Figure D.1).

It should be also emphasized, that these bond-projected measures of the AO overlaps in molecular bond system, which exhibit several properties and relations of ordinary probabilities, may assume negative values. Nevertheless, these generalized, projection measures obey several relations of the ordinary probabilities, e.g., the relevant sum rules, the relation between conditional couplings implied by the alternative expressions for the simultaneous interactions of three-AO in molecular bond system,

$$Q(i\wedge j\wedge k) = p_i\, Q(j\wedge k|i) = P(i\wedge j)\, Q(k|i\wedge j),\tag{D.6}$$

where $Q(k|i\wedge j) = Q(i\wedge j\wedge k)/P(i\wedge j)$.

In a similar manner one generates the bond projected measures of the joint interactions of *four* AO in the molecule and their associated conditional couplings,

$$\mathbf{Q}(\boldsymbol{a}\wedge\boldsymbol{b}\wedge\boldsymbol{c}\wedge\boldsymbol{d}) = \{Q(i\wedge j\wedge k\wedge l) = p_l\, Q(i\wedge j\wedge k|l)\}, \tag{D.7}$$

$$\mathbf{Q}(\boldsymbol{a}\wedge\boldsymbol{b}\wedge\boldsymbol{c}\wedge\boldsymbol{d}) = \{Q(i\wedge j\wedge k\wedge l) = (8N)^{-1}\gamma_{i,j}\gamma_{j,k}\gamma_{k,l}\gamma_{l,i}$$

$$= (8N)^{-1}\langle i|\hat{\mathrm{P}}_\varphi|j\rangle\langle j|\hat{\mathrm{P}}_\varphi|k\rangle\langle k|\hat{\mathrm{P}}_\varphi|l\rangle\langle l|\hat{\mathrm{P}}_\varphi|i\rangle\}, \tag{D.8}$$

$$\mathbf{Q}(\boldsymbol{a}\wedge\boldsymbol{b}\wedge\boldsymbol{c}|\boldsymbol{d}) = \{Q(i\wedge j\wedge k|l) = Q(i\wedge j\wedge k\wedge l)/p_l = (8\gamma_{l,l})^{-1}\gamma_{i,j}\gamma_{j,k}\gamma_{k,l}\gamma_{l,i}\}. \tag{D.9}$$

Again, these joint overlaps satisfy the required sum rules, e.g.,

$$\Sigma_i\Sigma_j\Sigma_k\, Q(i\wedge j\wedge k|l) = 1,\ \Sigma_l\, Q(i\wedge j\wedge k\wedge l) = Q(i\wedge j\wedge k),\ \text{etc.}, \tag{D.10}$$

but – similarly to the *three*-AO indices of Eqs. (C.4) and (C.5) – may assume negative values (see Figure D.1). Therefore, when taking the logarithm a care should be taken to use their moduli.

The fulfilment of the normalization relations of the preceding equation is again guaranteed by the idempotency condition of Eq. (199). For example,

$$\Sigma_i\Sigma_j\Sigma_k\, Q(i\wedge j\wedge k|l) = (8\gamma_{l,l})^{-1}\,\Sigma_i\Sigma_j\,(\Sigma_k\,\gamma_{j,k}\gamma_{k,l})\gamma_{i,j}\,\gamma_{l,i}$$

$$= (4\gamma_{l,l})^{-1}\,\Sigma_i\,(\Sigma_j\gamma_{i,j}\gamma_{j,l})\gamma_{l,i}$$

$$= (2\gamma_{l,l})^{-1}\,(\Sigma_i\gamma_{l,i}\gamma_{i,l}) = \gamma_{l,l}/\gamma_{l,l} = 1, \tag{D.11}$$

$$\Sigma_l\, Q(i\wedge j\wedge k\wedge l) = (8N)^{-1}(\Sigma_l\gamma_{k,l}\gamma_{l,i})\gamma_{i,j}\gamma_{j,k} = (4N)^{-1}\gamma_{k,i}\gamma_{i,j}\gamma_{j,k} = Q(i\wedge j\wedge k). \tag{D.12}$$

This development can be straightforwardly extended to cover a general case of the n-orbital couplings $\mathbf{Q}(\boldsymbol{a}_1\wedge\boldsymbol{a}_2\wedge\ldots\wedge\boldsymbol{a}_n) = \{Q(i_1\wedge i_2\wedge\ldots\wedge i_n)\}$ and $\mathbf{Q}(\boldsymbol{a}_1\wedge\boldsymbol{a}_2\wedge\ \ldots\ \wedge\boldsymbol{a}_{n-1}|\boldsymbol{a}_n) = \{Q(i_1\wedge i_2\wedge\ldots\wedge i_{n-1}|i_n) = Q(i_1\wedge i_2\wedge\ldots\wedge i_n)/P(i_n)\}$[20,22]:

$$\mathbf{Q}(\boldsymbol{a}_1\wedge\boldsymbol{a}_2\wedge\ldots\wedge\boldsymbol{a}_n) = \{Q(i_1\wedge i_2\wedge\ldots\wedge i_n) = (2^{n-1}N)^{-1}\,\gamma_{i_1,i_2}\gamma_{i_2,i_3}\cdots\gamma_{i_{n-1},i_n}\gamma_{i_n,i_1}$$

$$= (2^{n-1}N)^{-1}\langle i_1|\hat{\mathrm{P}}_\varphi|i_2\rangle\langle i_2|\hat{\mathrm{P}}_\varphi|i_3\rangle\ \ldots\ \langle i_{n-1}|\hat{\mathrm{P}}_\varphi|i_n\rangle\langle i_n|\hat{\mathrm{P}}_\varphi|i_1\rangle\}, \tag{D.13}$$

$$\mathbf{Q}(\boldsymbol{a}_1\wedge\boldsymbol{a}_2\wedge\ldots\wedge\boldsymbol{a}_{n-1}|\boldsymbol{a}_n) = \{Q(i_1\wedge i_2\wedge\ldots\wedge i_{n-1}|i_n)$$

$$= (2^{n-1}\gamma_{i_n,i_n})^{-1}\,\gamma_{i_1,i_2}\gamma_{i_2,i_3}\cdots\gamma_{i_{n-1},i_n}\gamma_{i_n,i_1}\}. \tag{D.14}$$

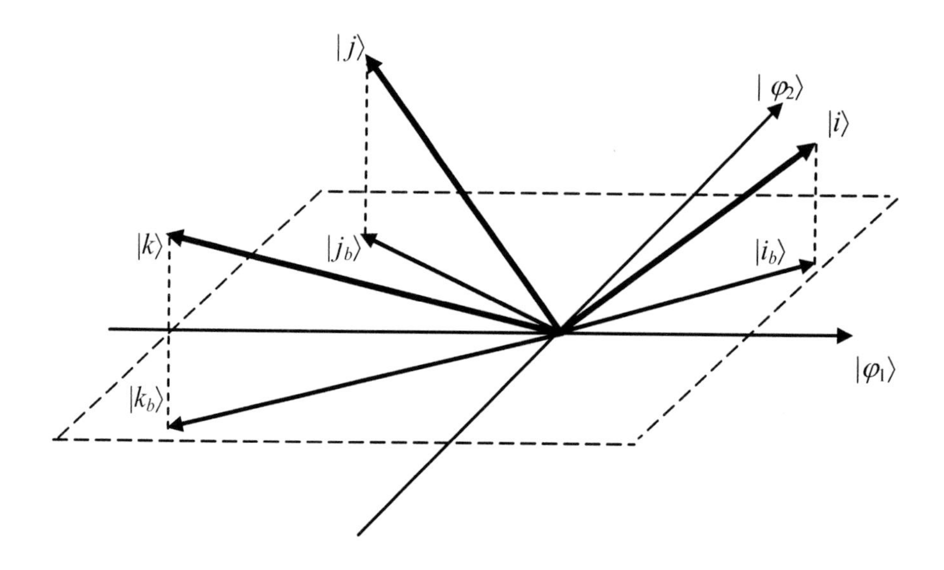

Figure D.1. The geometric interpretation of the projection $\{|i_b\rangle = \hat{\mathrm{P}}_\varphi|i\rangle \equiv \boldsymbol{i}_b \cdot |\boldsymbol{\varphi}\rangle = \sum_s i_s |\varphi_s\rangle,\ |j_b\rangle = \hat{\mathrm{P}}_\varphi|j\rangle,\ |k_b\rangle = \hat{\mathrm{P}}_\varphi|k\rangle\}$ of three AO participating in chemical bonds, $\{\chi_i = |i\rangle,\ \chi_j = |j\rangle,\ \chi_k = |k\rangle\}$, onto the subspace (plane) $\{\varphi_1 = |\varphi_1\rangle,\ \varphi_2 = |\varphi_2\rangle\}$ of the two occupied MO. These projections determine the bond orders, e.g., $\gamma_{i,j} = \langle i|\hat{\mathrm{P}}_\varphi|j\rangle = \boldsymbol{i}_b \cdot \boldsymbol{j}_b$, and hence the bond-projected measure of the joint interaction (overlap) $Q(i\wedge j\wedge k) = (4N)^{-1}\gamma_{i,j}\gamma_{j,k}\gamma_{k,i} = (2/N)\,(\boldsymbol{i}_b \cdot \boldsymbol{j}_b)\,(\boldsymbol{j}_b \cdot \boldsymbol{k}_b)\,(\boldsymbol{k}_b \cdot \boldsymbol{i}_b)$. It should be observed that the *off*-diagonal element $\gamma_{i,j}$ of the density matrix becomes negative, when the effective interaction between χ_i and χ_j in all occupied MO becomes *anti*-bonding, i.e., when the "angle" between the bond projections of these two AO $\alpha(\boldsymbol{i}_b, \boldsymbol{j}_b) > \pi/2$.

Yet another approach to *multi*-orbital probabilities uses the so called normalized *coupling-coefficient* as probability amplitude. Consider again the normalized *two*-AO bond projection,

$$q(i\wedge j) = \sqrt{\frac{2}{N}}\langle j_b|i_b\rangle = \sqrt{\frac{2}{N}}\langle j|\hat{\mathrm{P}}_\varphi|i\rangle, \tag{D.15}$$

measuring the common part of two AO in the occupied MO subspace,

$$\hat{\mathrm{P}}_j^b|i_b\rangle = |j_b\rangle\langle j_b|i_b\rangle, \tag{D.16}$$

which provides the amplitude of the joint probability of Eq. (204): $P(i\wedge j)$ = $|q(i\wedge j)|^2 \equiv p(i\wedge j)$. One similarly uses the common, bonding part of three AO,

$$\hat{P}_k^b \hat{P}_j^b |i_b\rangle = |k_b\rangle\langle k_b | j_b\rangle\langle j_b | i_b\rangle, \tag{D.17}$$

as the amplitude of the associated measure of the *three*-AO probability, in the spirit of the superposition principle of quantum mechanics (see Appendix C):

$$q(i\wedge j\wedge k) = \sqrt{\frac{2}{N}}\langle k_b | j_b\rangle\langle j_b | i_b\rangle = \sqrt{\frac{2}{N}}\langle k|\hat{P}_\varphi|j\rangle\langle j|\hat{P}_\varphi|i\rangle, \tag{D.18}$$

$$p(i\wedge j\wedge k) = |q(i\wedge j\wedge k)|^2 = \frac{2}{N}\langle i|\hat{P}_\varphi|j\rangle\langle j|\hat{P}_\varphi|k\rangle\langle k|\hat{P}_\varphi|j\rangle\langle j|\hat{P}_\varphi|i\rangle$$

$$= \frac{1}{8N}(\gamma_{i,j}\gamma_{j,i})(\gamma_{j,k}\gamma_{k,j}) \geq 0. \tag{D.19}$$

The partial normalization, of $q(i\wedge j\wedge k)$ to $q(i\wedge j)$, which then automatically guarantees the appropriate normalization of $p(i\wedge j\wedge k)$, follows from the completeness in the occupied-MO subspace of the bond-projected AO basis, $\sum_m \hat{P}_m^b = 1$,

$$\Sigma_k \hat{P}_k^b \hat{P}_j^b |i_b\rangle = \hat{P}_j^b |i_b\rangle = |j_b\rangle\langle j_b | i_b\rangle$$

or

$$\Sigma_k |k_b\rangle\, q(i\wedge j\wedge k) = \sqrt{\frac{2}{N}}\sum_k |k_b\rangle\langle k_b | j_b\rangle\langle j_b | i_b\rangle = \sqrt{\frac{2}{N}}\,|j_b\rangle\langle j_b | i_b\rangle = |j_b\rangle q(i\wedge j). \tag{D.20}$$

The associated *four*-AO amplitude,

$$q(i\wedge j\wedge k\wedge l) = \sqrt{\frac{2}{N}}\langle l_b | k_b\rangle\langle k_b | j_b\rangle\langle j_b | i_b\rangle = \sqrt{\frac{2}{N}}\langle l|\hat{P}_\varphi|k\rangle\langle k|\hat{P}_\varphi|j\rangle\langle j|\hat{P}_\varphi|i\rangle, \tag{D.21}$$

which generates the joint probability $P(i\wedge j\wedge k\wedge l)$ of the simultaneous four-AO events in the occupied MO subspace,

$$p(i \wedge j \wedge k \wedge l) = |q(i \wedge j \wedge k \wedge l)|^2$$

$$= \frac{2}{N}\langle i|\hat{\mathrm{P}}_\varphi|j\rangle\langle j|\hat{\mathrm{P}}_\varphi|k\rangle\langle k|\hat{\mathrm{P}}_\varphi|l\rangle\langle l|\hat{\mathrm{P}}_\varphi|k\rangle\langle k|\hat{\mathrm{P}}_\varphi|j\rangle\langle j|\hat{\mathrm{P}}_\varphi|i\rangle \geq 0, \qquad \text{(D.22)}$$

similarly follows from the common part of the four AO in φ:

$$\hat{\mathrm{P}}_l^b\hat{\mathrm{P}}_k^b\hat{\mathrm{P}}_j^b|i_b\rangle = |l_b\rangle\langle l_b|k_b\rangle\langle k_b|j_b\rangle\langle j_b|i_b\rangle. \qquad \text{(D.23)}$$

The partial normalization of $q(i \wedge j \wedge k \wedge l)$ to $q(i \wedge j \wedge k)$ again follows from the identity holding in the occupied MO subspace, $\sum_l \hat{\mathrm{P}}_l^b = 1$:

$$\Sigma_l |l_b\rangle q(i \wedge j \wedge k \wedge l) = \sqrt{\frac{2}{N}}\sum_l |l_b\rangle\langle l_b|k_b\rangle\langle k_b|j_b\rangle\langle j_b|i_b\rangle$$

$$= \sqrt{\frac{2}{N}}\,|k_b\rangle\langle k_b|j_b\rangle\langle j_b|i_b\rangle = |k_b\rangle q(i \wedge j \wedge k). \qquad \text{(D.24)}$$

It should be observed that the *joint*-probability measures of Eqs. (D.19) and (D.22) are not symmetrical with respect to exchanging the AO indices. In a sense they preserve a memory of the sequence of AO projections:

$$p(i \wedge j \wedge k) = P(i \,\vert\, j \,\vert\, k), \quad p(i \wedge j \wedge k \wedge l) = P(i \,\vert\, j \,\vert\, k \,\vert\, l), \text{ etc.} \qquad \text{(D.25)}$$

Here, the broken vertical line marks the projection dependence of the right AO on the left AO in the probability symbol. This development can be straightforwardly generalized for any number of AO in the sequence.

The true measures of the *joint* AO probabilities $P(i \wedge j \wedge k)$ and $P(i \wedge j \wedge k \wedge l)$ can be then obtained by the symmetrization over all $n!$ permutations $\Pi(i_1 \wedge i_2 \wedge \ldots \wedge i_n) = (i_1^\Pi, i_2^\Pi, \ldots, i_n^\Pi)$ over the n AO indices involved [20]:

$$P(i_1 \wedge i_2 \wedge \ldots \wedge i_n) = \frac{1}{n!}\sum_\Pi p(i_1^\Pi \wedge i_2^\Pi \wedge \ldots \wedge i_n^\Pi). \qquad \text{(D.26)}$$

ACKNOWLEDGMENTS

I wish to express my thanks to Dr. E. Broniatowska, for preparing Figs. 2-11 and Tables 1-3, and to Dr. S. Escalante – for drawing Figs. 12-14. Thanks are also due to Mr. P. de Silva, who has prepared Figs. 17-25, and Mr. D. Szczepanik, who has generated the RHF OCT results of Tables 4 and 5.

BIBLIOGRAPHY

[1] R. A. Fisher, Proc. Cambridge Phil. Soc. 22, 700 (1925).

[2] B. R. Frieden, *Physics from the Fisher Information – A Unification*, 2nd Ed. (Cambridge University Press, Cambridge, 2004).

[3] C. E. Shannon, Bell System Tech. J. 27, 379, 623 (1948); C. E. Shannon, W. Weaver, *The Mathematical Theory of Communication*, (University of Illinois, Urbana, 1949).

[4] S. Kullback, R. A. Leibler, Ann. Math. Stat. 22, 79 (1951).

[5] S. Kullback, *Information Theory and Statistics* (Wiley, New York, 1959).

[6] N. Abramson, *Information Theory and Coding*, (McGraw-Hill, New York, 1963); see also: P. E. Pfeifer, *Concepts of Probability Theory*, 2nd Ed. (Dover, New York, 1978).

[7] R. F. Nalewajski, *Information Theory of Molecular Systems* (Elsevier, Amsterdam, 2006), and refs. therein.

[8] R. F. Nalewajski, (a) Adv. Quant. Chem. 56, 217 (2009); (b) R. F. Nalewajski, D. Szczepanik, J. Mrozek, Adv. Quant. Chem., in press.

[9] R. F. Nalewajski, E. Świtka, A. Michalak, Int. J. Quantum. Chem. 87, 198 (2002); R. F. Nalewajski, E. Świtka, Phys. Chem. Chem. Phys. 4, 4952 (2002).

[10] R. F. Nalewajski and E. Broniatowska, J. Phys. Chem. A. 107, 6270 (2003).

[11] R. F. Nalewajski and R. G. Parr, Proc. Natl. Acad. Sci. USA 97, 8879 (2000); J. Phys. Chem. A 105, 7391 (2001).

[12] R. F. Nalewajski and R. Loska, Theoret. Chem. Acc. 105, 374 (2001).

[13] R. F. Nalewajski, Phys. Chem. Chem. Phys. 4, 1710 (2002).

[14] R. F. Nalewajski, Chem. Phys. Lett. 372, 28 (2003).

[15] R. G. Parr, P. W. Ayers, R. F. Nalewajski, J. Phys. Chem. A 109, 3957 (2005).

[16] R. F. Nalewajski, Adv. Quant. Chem. 43, 119 (2003).

[17] R. F. Nalewajski and E. Broniatowska, Theoret. Chem. Acc. 117, 7 (2007).
[18] R. F. Nalewajski, (a) J. Math. Chem. 43, 265 (2008); (b) *Ibid.*, 44, 414 (2008); (c) *Ibid.*, 45, 607 (2009); (d) *Ibid.*, 45, 709 (2009); (e) *Ibid.*, 45, 776 (2009); (f) *Ibid.*, 45, 1041, (2009).
[19] R. F. Nalewajski, J. Math. Chem. 47, 667 (2010).
[20] R. F. Nalewajski, J. Math. Chem. 47, 692 (2010).
[21] R. F. Nalewajski, J. Math. Chem. 47, 709 (2010).
[22] R. F. Nalewajski, J. Math. Chem. 47, 808 (2010).
[23] F. L. Hirshfeld, Theoret. Chim. Acta (Berl.) 44, 129 (1977).
[24] R. F. Nalewajski, Int. J. Quantum Chem. 108, 2230 (2008).
[25] (a) R. F. Nalewajski, P. de Silva, J. Mrozek, in: *Kinetic Energy Functional*, edited by A. Wang and T. Wesołowski (World Scientific, Singapore, 2009), in press; (b) *Ibid.*, THEOCHEM, Proceedings of 13th Conference on the Applications of Density Functional Theory in Chemistry and Physics, Lyon 31st August-4th September 2009, J.-L. Rivail, Ed., in press.
[26] R. F. Nalewajski, A. M. Köster, S. Escalante, J. Phys. Chem. A 109, 10038 (2005).
[27] A. D. Becke, K. E. Edgecombe, J. Chem. Phys. 92, 5397 (1990).
[28] R. F. Nalewajski, J. Phys. Chem. A 104, 11940 (2000).
[29] R. F. Nalewajski, K. Jug, in: *Reviews of Modern Quantum Chemistry: A Celebration of the Contributions of Robert G. Parr*, edited by K. D. Sen, (World Scientific, Singapore, 2002), Vol. I, p. 148; R. F. Nalewajski, Struct. Chem. 15, 391 (2004).
[30] R. F. Nalewajski, Mol. Phys. 102, 531, 547 (2004); *Ibid.*, Mol. Phys. 103, 451 (2005); *Ibid.* 104, 365, 493, 1977, 2533 (2006).
[31] R. F. Nalewajski, Theoret. Chem. Acc. 114, 4 (2005); *Ibid.*, J. Math. Chem. 38, 43 (2005).
[32] R. F. Nalewajski, Int. J. Quantum Chem. 109, 425, 2495 (2009).
[33] R. F. Nalewajski, (a) J. Math. Chem. 43, 780 (2008); (b) J. Phys. Chem. A 111, 4855 (2007).
[34] R. F. Nalewajski, (a) Mol. Phys. 104, 3339 (2006); (b) Adv. Quant. Chem., 56, 217 (2009).
[35] R. F. Nalewajski, J. Phys. Chem A107, 3792 (2003); *Ibid.*, Mol. Phys. 104, 255 (2006); *Ibid.*, Ann. Phys. (Leipzig) 13, 201 (2004).
[36] C. F. von Weizsäcker, Z. Phys. 96, 431 (1935).
[37] S. B. Sears, Ph. D. Thesis, The University of North Carolina at Chapel Hill (1980); S. B. Sears, R. G. Parr, U. Dinur, Israel J. Chem. 19, 165 (1980).

[38] B. R. Frieden, Am. J. Phys. 57, 1004 (1989); M. Reginatto, Phys. Rev. A58, 1775 (1998); Erratum: Phys. Rev. A60, 1730 (1999).
[39] K. A. Wiberg, Tetrahedron 24, 1083 (1968).
[40] W. Kohn, L.J. Sham, Phys. Rev. 140A, 1133 (1965).
[41] P. Hohenberg, W. Kohn, Phys. Rev. 136B, 864 (1964).
[42] R. G. Parr, W. Yang, *Density-Functional Theory of Atoms and Molecules* (Oxford University Press, New York, 1989).
[43] R. M. Dreizler, E. K. U. Gross, *Density Functional Theory: An Approach to the Quantum Many-Body Problem* (Springer-Verlag, 1990).
[44] R. F. Nalewajski, Ed., *Density Functional Theory I-IV*, Topics in Current Chemistry, Vols.180-183 (Springer-Verlag, Berlin, 1996).
[45] P. A. M. Dirac, *The Principles of Quantum Mechanics*, 4th ed. (Clarendon, Oxford, 1958).
[46] R. F. Nalewajski, A. M. Köster, K. Jug, Theoret. Chim. Acta (Berl.) 85, 463 (1993).
[47] R. F. Nalewajski, J. Mrozek, Int. J. Quantum Chem. 51, 187 (1994).
[48] R. F. Nalewajski, S. J. Formosinho, A. J. C. Varandas, J. Mrozek, Int. J. Quantum Chem. 52, 1153 (1994).
[49] R. F. Nalewajski, J. Mrozek, G. Mazur, Can. J. Chem. 100, 1121 (1996).
[50] R. F. Nalewajski, J. Mrozek, A. Michalak, Int. J. Quantum Chem. 61, 589 (1997).
[51] J. Mrozek, R. F. Nalewajski, A. Michalak, Polish J. Chem. 72, 1779 (1998).
[52] R. F. Nalewajski, Chem. Phys. Lett. 386, 265 (2004).
[53] M. S. Gopinathan, K. Jug, Theor. Chim. Acta (Berl.) 63 497, 511 (1983); see also: K. Jug. M. S. Gopinathan, in *Theoretical Models of Chemical Bonding*, edited by Z. B. Maksić, (Springer, Heidelberg, 1990), Vol. II, p. 77.
[54] I. Mayer, Chem. Phys. Lett. 97, 270 (1983).
[55] R. F. W. Bader, *Atoms in Molecules* (Oxford University Press, Oxford, 1990).
[56] J. F. Capitani, R. F. Nalewajski and R. G. Parr, J. Chem. Phys. 76, 568 (1982).
[57] R. G. Gordon and Y. S. Kim. J. Chem. Phys. 56, 3122 (1972).
[58] P. Cortona, Phys. Rev. B 44, 8454 (1991).
[59] T. Wesołowski and A. Warshel, J. Phys. Chem. 97, 8050 (1993).
[60] T. Wesołowski, J. Am. Chem. Soc. 126, 11444 (2004); *Ibid.*, Chimia 58, 311 (2004), and refs. therein.

[61] B. Silvi and A. Savin, Nature 371, 683 (1994).
[62] A. Savin, R. Nesper, S. Wengert and T. F. Fässler, Angew. Chem. Int. Ed. Engl. 36, 1808 (1997).
[63] M. J. Feinberg, K. Ruedenberg and E. L. Mehler, Adv. Quant. Chem. 5, 28 (1970).
[64] M. J. Feinberg and K. Ruedenberg, J. Chem. Phys. 54, 1495 (1971); *Ibid.* 55, 5804 (1971).
[65] W. A. Goddard and C. W. Wilson, Theoret. Chim. Acta (Berl.) 26, 195, 211 (1972).
[66] R. G. Pearson, J. Am. Chem. Soc. 85, 3533 (1963); Science 151, 172 (1966).
[67] R. G. Pearson, *Hard and Soft Acids and Bases* (Dowden, Hatchinson and Ross, Stroudsburg, 1973).
[68] C. K. Jørgensen, Inorg. Chem. 3, 1201 (1964).
[69] R. G. Parr and R. G. Pearson, J. Am. Chem. Soc. 105, 7512 (1983); see also: R. F. Nalewajski, J. Am. Chem. Soc. 106, 944 (1984).
[70] R. F. Nalewajski, J. Korchowiec and A. Michalak, Topics in Current Chemistry, 183, 25 (1996).
[71] R. F. Nalewajski and J. Korchowiec, *Charge Sensitivity Approach to Electronic Structure and Chemical Reactivity* (World-Scientific, ingapore,1997).
[72] R. F. Nalewajski, Adv. Quant. Chem. 51, 235 (2006).
[73] A. K. Chandra, A. Michalak, M. T. Nguyen and R. F. Nalewajski, J. Phys. Chem. A 102, 10182 (1998).
[74] R. F. Nalewajski, Topics in Catalysis 11/12, 469 (2000).
[75] V. Gutmann, *The Donor Acceptor Approach to Molecular Interactions* (Plenum, New York, 1978).
[76] R. F. Nalewajski, J. Math. Chem. 44, 802 (2008).
[77] M. Mitoraj and A. Michalak, J. Mol. Model. 11, 341 (2005); *Ibid.*, 13, 347 (2007).
[78] M. Mitoraj, Ph. D. Thesis, Jagiellonian University, 2007.
[79] M. Mitoraj, H. Zhu, A. Michalak, and T. Ziegler, J. Org. Chem. 71, 9208 (2006); *Ibid.*, Organometallics 26, 1627 (2007).
[80] E. Runge and E. K. U. Gross, Phys. Rev. Lett. 52, 997 (1984).
[81] E. K. U. Gross, J. F. Dobson and M. Petersilka, Topics in Current Chemistry 181, 81 (1996).
[82] H. B. Callen, Thermodynamics: an Introduction to the physical Theories of Equilibrium Thermostatics and Irreversible Thermodynamics (Wiley, New York, 1960).

INDEX